HOW TOs

This book features an array of articles to help you perform your own modifications. It doesn't cover everything, though, and you'll still need to visit a specialist for some things, but we'll show you how to perform some of the jobs you probably thought were too hard.

Experience has shown that even when you pay top dollar for a job, there are businesses out there that are willing to take your hard-earned cash for merely tightening a component or reattaching a wire. Not all businesses operate in this manner, mind you, but a few do. Hopefully, this book will help you eliminate the problem and keep your cash for more important modifications.

In the following pages there are plenty of tips on do-it-yourself modifications, such as fitting door locks, steering basics, installing cruise control and tinting windows. It's all in here. So sit back, study each and every page, then get ready to get your hands dirty. It's never as hard as it seems.

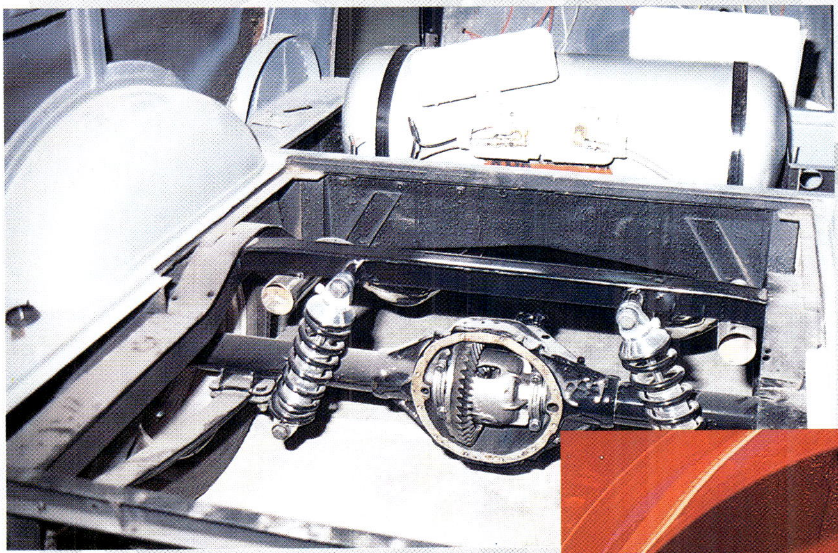

P E R F O R M A N C E   C A R

HOW TOS

Performance Car How Tos is published by Clockwork Media Pty Ltd, Suite 4, 4th Floor Randall House, 117-119 Pacific Highway, Hornsby, New South Wales 2077 Australia. PO Box 5277 Erina Fair, New South Wales 2250 Australia. Australia: Phone (02) 9477 4300. Fax (02) 9477 4344. International: Phone 61 2 9477 4300. Fax 61 2 9477 4344. Printing by Times offset (M) Sdn, Bhd, Malaysia. Distribution enquiries to (02) 9805 0399.

ISBN 0 947216 75 8

ACN 003 609 103

DO YOUR OWN MACHINING? Now you can.

Story/photos by Paul Tuszon

The thing that pushes most of us to build our own cars is probably the money we save by doing it ourselves. Of course, there's also the satisfaction of having completed the job with our own hands.

Saving money and feeling satisfied in this way has been possible in many areas of building your dream machine. Bodywork, suspension and fitting external components like exhausts and carburetors have always been realistic objectives for the D-I-Y car-builder. And those with more experience can even assemble their own engines successfully.

DOING THE IMPOSSIBLE

However, one area that has always remained impossible for the home mechanic has been the actual machining of engine components. Even if you did have the tens of thousands of dollars worth of equipment needed, you still couldn't do it without training.

Well, now both the training and equipment are available to you. The Barton College of TAFE is running a course in which you can learn to use all the machines needed to recondition your motor. This is welcome news for any enthusiast wanting greater knowledge, but the best part is that you get to take your own motor along and work on it as you learn.

Each class has six to 10 students and one instructor. Each session runs for three hours on Wednesday evenings starting at 6pm. There's no set curriculum; you simply tell the instructor what you want to do and he shows you how. This is economical because you don't spend time learning things you don't want to know. On the other hand, the individual attention required by this method means the instructor may not always be on hand right when you want him. The best way of doing things is to make sure you have some other work you can go on with while you wait for him.

TAKEAWAY ENGINES

There isn't much storage space available, so students have to take their parts home with them after each class. This isn't as bad as it sounds because you can drive right into the workshop and park under the overhead crane. Also, the atmosphere is pretty friendly and there's always someone walking past who'll lend a hand to lift a block or whatever.

A fee of $336 buys you 10 Wednesday nights. Realistically speaking, you probably won't get a whole engine done in this time; I'm doing the course now and I know I won't. However, even if you have to enrol twice you'll still be in front when you consider the cost of custom engine machinework. Use of all equipment is included in the fee but, obviously, you'll have to supply all your own parts.

To help wring every last dollar's worth of value from the course, you shouldn't go along to the first session empty-handed wondering what's going to happen — take an engine and get stuck into it straight away! The disassembly area and hot-tank facilities are excellent. It would be even more economical to disassemble your engine at home and drop the parts in the hot tank as soon as you arrive. That means they'll be getting cleaned as you take the fairly short introductory tour on your first evening.

WHATEVER YOU NEED

Almost every type of engine-building machine is available at the course. In addition to those shown in the photographs, there are presses, MIG and oxy welders, lathes, an excellent engine assembly area, an engine dyno for first starts, and all sorts of other small pieces of equipment such as grinders, air tools, micrometers, etc. There's also a large cracktesting machine that's capable of taking a block. It's so easy to use you'll want to test everything, and so you should.

This is an excellent course which enables you to do things that once would have been impossible for a non-professional. Completing it means you really will be able to say you built your car yourself. You can reach the Barton College of TAFE on (03) 9209 5000. Of course, Barton's not the only TAFE college in the country; they're all over the place. We're not sure how many of them have similar facilities to those at Barton but ring the college closest to you and find out — it's well worth the effort.

The boring bar is large and capable of taking any motor for a road-going vehicle. The block in the shot is a Jaguar 60-degree V12. Roger, the head instructor, loves Jags.

The Sunnen honing machine is the newest machine at the college. It's completely automated and avoids the inconsistencies that honing by hand can produce. This is particularly important with an inexperienced operator.

An accurate main bearing tunnel is essential in even a stock engine. If you are planning to boost power beyond standard (and who isn't) then it's even more important.

Complete head preparation facilities are available including all the air-tools needed for jobs ranging from basic reconditioning to full blown porting. This is a valveseat cutter.

Accurate seats aren't much use without accurate valvefaces.

Headwork isn't much good if it doesn't improve flow and the only way to find out if it does is with a proper flowbench.

Big ends on conrods which have become oval due to wear or which have been distorted by fitting new, higher-quality bolts will need to be resized.

The college has this rod-boring machine which can be used to convert rods from pressed-pin to full-floating.

Engine balance is critical and now you can do your own. This machine will also do transmission driveshafts.

Crank bearing clearances are only about two to three thousandths of an inch. If your crank has grooves deeper than this it will need to be reground.

Heads must be flat for a number of reasons, so no workshop should be without a decent-sized milling machine.

You can even grind a camshaft to your own specifications if you want to.

The disassembly and hot-tank cleaning facilities are really good. You'll get your components cleaner than you ever could at home.

All the equipment needed for accurate valve jobs is here. Up to four people can work on the head preparation benches at one time.

Fitting Timing Gears to your engine

Timing gears are a crucial factor in achieving peak performance from your engine. Follow this step-by-step guide to installing a new timing gear set and enjoy the instant rewards

CAM DRIVE SYSTEMS

Many standard engines are fitted with a single-row timing chain and cast timing gears which are not suitable for the higher loads generated by performance cam and spring combinations and the higher rev ranges that result.

The next step up the performance ladder in camshaft drive systems is the double-row roller chain and steel gear combination. These are available for most popular engines and offer excellent durability and accurate cam timing in all but the most radical applications.

A number of these sets feature multiple keyways on the crank gear to allow easy cam timing adjustment. Some manufacturers also harden the teeth of the steel crank gears to offer even greater durability on this fast-spinning gear.

While the double-row timing chain sets are fine for most performance applications, the chain still represents the weak link and can suffer stretch and even breakage under ultra-high loads and rpm.

At the top of the ladder is the gear drive. These are used to eliminate the chain and offer even greater durability under the most severe performance demands. It must be remembered, however, that gear drives are not known for quiet running — though to some people that's a positive.

When the chain system in your engine has worn or stretched is a good time to consider updating your cam drive system to either a double-row chain and steel gear set or going all the way to peak performance and timing accuracy by installing a gear drive set.

It's a job that can be carried out while the engine is still in the car but it will require the removal of the radiator and possibly the grille to allow access to the work area.

You remove and replace the same components as you would in changing a camshaft. If you are considering changing your camshaft you may well find that your existing chain set is worn and will need replacement, too. This would be the ideal time to upgrade your cam drive system.

ext and pix by Rob Warden

We chose an Australian-made cam gear drive set for several reasons. Not only is the quality as good as or better than in imported systems, but all parts are readily available with local product support.

Crow Cams offers a range of cam drive systems to suit most popular local cars and also carries a complete valvetrain component range to suit every car and application. The Crow Cams computer camshaft selection service, free to all customers, ascertains which are the right products for the application.

For this feature we've used a Holden V8 typical of thousands found on Aussie roads, but the principles are the same for most engines. Each timing gear set includes installation instructions but these photos are a hands-on guide to make it all a little clearer.

SOURCE

Crow Cams
16 Colbert Road, Campbellfield, Vic 3061
Phone (03) 357 0469,
fax (03) 357 0001

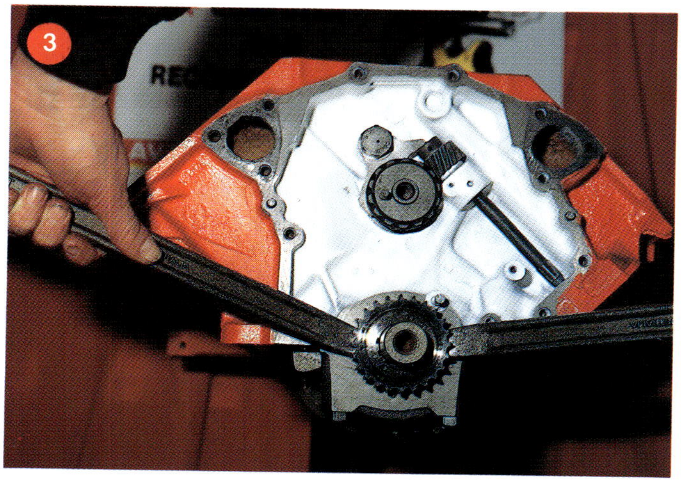

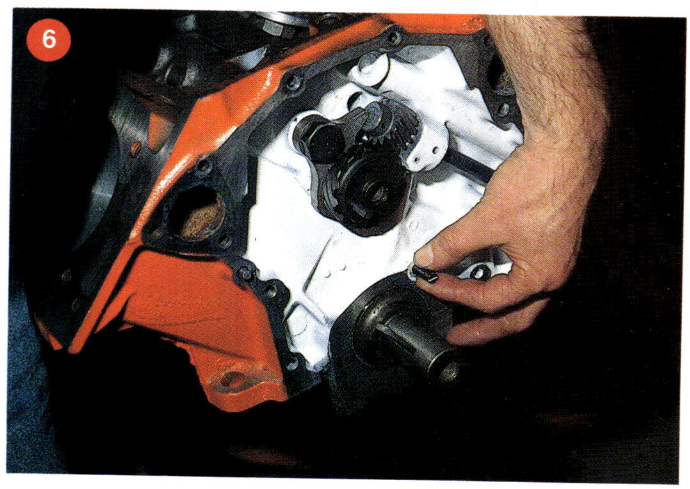

7

8

9

10

11

12

1. Once you have removed the radiator, water pump, harmonic balancer and timing case this is what you'd expect to see — though it probably wouldn't be as clean as this and fitted with a Crow Cams double-row timing chain and gear set. This is an aftermarket performance chain and gear set with a nine-option keyway bottom sprocket for complete cam timing adjustment. The bottom sprocket is induction-hardened and the cam gear is of lightweight machined steel.

2. Remove the cam gear and chain by removing the centre bolt and fuel pump eccentric.

3. The crank gear may need some leverage to remove it.

4. The Crow Cams performance timing gear kits are available to suit most popular engines. The kit comes with instructions.

5. Ensure that the engine is at top dead centre on No 1 cylinder during the assembly process. Remove the original chain oiler using vice grips.

6. Line up the new oiler (as supplied with kit) with the line horizontal and the hole towards the idler gear position.

7. Tap into hole gently, retaining correct alignment.

8. Oiler must protrude 8mm from block, no more or less.

9. When fitting crank gear ensure that the chamfer is closest to the block.

10. Line up the keyway and tap

▶ gently till firm against the crank.

11. On some blocks the casting around the oil pump drive boss is too high and may interfere with movement of the cam gear.

12. If there are clearance problems, file until casting no longer interferes with cam gear. Ensure that no filings contaminate the engine. Protect with rags and the like. Fit the cam (top) gear into position.

13. Slide the idler gears into position between the main gears. The larger idler gear must be fitted with the long part of the shaft towards the block.

14. Note that the larger idler gear is on the drive side. With the cam gear bolted up tight, ensure that the idler gears have 20-60 thou vertical clearance.

15. Be sure that both dots on the main gears are lined up when the engine is at TDC before final tightening of gears.

16.&17. On a Holden V8 this lug has to be removed from the inside of the timing cover to allow the idler gears to float freely between the block and the timing cover. Grind off the lug carefully without damaging any other parts of the cover. (Note: some aftermarket timing covers may require extra metal removal in this area.)

18. Trial-fit the timing cover after removing the lug. Do not use a gasket at this stage and make sure timing cover meets flush at all edges

with the block. This is so that the idler gears will have sufficient space when the cover is finally fitted along with the gasket.

19. Check that the engine is at top dead centre. Dial in camshaft to correct specifications using one of the five offset bushes supplied in the kit. These bushes give a range of adjustment from eight degrees retard to eight degrees advance.

20. Insert the required bush.

21. Care must be taken to hold the drive idler gear in to mesh with the main gears to eliminate backlash while dialling the cam in.

22. Fit the fuel pump eccentric and cam retaining bolt; tension to specs. Reassemble the front of the engine using new gaskets.

23. The new Crow Cams performance timing gears are in place and ready for final engine assembly and improved performance.

CROW CAMS

How To

Getting a Good Head Job

Because most power gains will be found in the heads, it follows that poor-quality headwork can be responsible for the failure of an engine to live up to its potential for performance.

By Paul Tuzson

Getting the gases flowing to maximum advantage within the combustion chamber is vital to any performance engine. Buying the best quality valves, rockers, guide-plates, pushrods and other hard parts is a good start, but if the standard of work that goes into preparing the head is not up to scratch, your money will be wasted.

Manufacturers' tolerances can often be improved upon outside the mass production environment because of the extra care a skilled machinist can devote to a job. However, valve seat concentricity is one area in which it is difficult to improve on factory specifications. Valve seat concentricity is the measurement of how accurately centred the valve seat is cut in relation to the valve guide.

The valve guides and seats are cut in one action during the original manufacturing operation. This means that concentricity from the factory is very accurate. If you check your shop manual you'll see that the recommended specifications are quite tight. Ford, for instance, specifies 0.0015 for a Cleveland. The question is, will your heads match these specifications after reconditioning?

Runout Problems

We'd heard that people were having trouble getting good work done in this area and just recently we encountered a pair of heads on which the owner had spent good money. Unfortunately for the reconditioner, this person had access to a valve seat run-out gauge.

He checked the work, for which he had paid lots of readies, only to discover that concentricity on some seats was out by up to 0.010 and that 0.005 was the closest any seat came to true concentricity.

The first step in getting a seat right is to get the valve guide right.

How could a manufacturer let such poor work leave the shop? Part of the reason is that valve seat runout gauges aren't often used to determine the accuracy of the work performed. A quicker and more common method is a vacuum tester.

While vacuum testers are not a bad idea, a concentricity problem has to be pretty severe before this method will pick it up. To demonstrate this point we enlisted the help of Paul Brandt at Eastern Cylinder Heads.

We asked Paul to prepare two chambers, with concentricity in one to be within tolerance, while we requested that the other be inaccurate enough to cause problems for an engine, yet still be able to pass a vacuum test.

Getting the Preliminaries Right

Paul says that the first step in getting a seat right is to get the valve guide right. With care this is usually fairly easy for a skilled operator, but in certain severe circumstances a guide could be too far gone for normal methods.

Rocker geometry is important here. If you have the wrong setup, your valves won't be simply pushed up and down in the guides, the rocker arm will try to push or pull them to one side. Eventually the guide will be worn oval at either end.

This problem is not difficult to understand when you consider that half an inch of valve lift means that at 5000rpm each valve-stem slides over one metre in its guide every second. The condition can become so pronounced that a bronze guide insert won't be able to make contact along the entire

2

3

length of the original guide. This situation will affect the permanence of the installation. Guides worn to this point need to be drilled out wider and fitted with thin-walled liners, which are actually thicker than bronze guide inserts.

4

The first step is to drill out the guides using special locating cones that keep the drill centralised in relation to the original seat. These come in different sizes. Care is needed for this step because it sets the standards for the operations which follow (photographs 2 & 3).

Any debris remaining in the guide after drilling will affect the fitting of the insert, so thorough cleaning with an appropriate brush is essential (photograph 4). The insert will slip into the guide a little easier with the application of some oil (photograph 5).

5

Insert Here

Most guides are not worn to the point where they need thin-walled inserts. Bronze guide inserts almost always do the job. Not everyone may be clear about the differences between thin-walled inserts and bronze guide inserts. Those shown in the photograph 6 are bronze guide inserts. As we mentioned before, thin-walled inserts are actually thicker than these, but they picked up their name before bronze guide inserts were available.

The brand name of the inserts pictured is K-Line Interrupted Spiral Bronze Guide-Liners. There are other perfectly good brands available, but these are so widely used that people often just use the term 'K-Lines' when referring to all bronze guide inserts.

The inserts are fitted into the installing sleeve as shown in photograph 6 and pushed into the bored-out guides with a pneumatic hammer (photograph 7).

The installing sleeve ensures that the ends of the inserts are tapped in flush with the ends of the guides that extend into the ports. At the rocker end of the guides the inserts

6

7

need to be trimmed off with this special little hand-operated tool (photograph 8).

It's important that each valve stem be perfectly aligned with the sides of the insert along its entire length. This assists with heat dissipation and helps prevent engine oil

8

leaking down into the chambers. It also guards against premature wear.

After the inserts are first fitted, the internal surfaces on which the valves must slide are quite irregular and show many high spots. Removal of these high spots by honing provides the best surface for valves to slide in, but it takes quite a long time.

Broaching the Subject

One method of speeding up the operation is by broaching the insert. This consists of using the pneumatic hammer to force ball broaches through the installed inserts. Such broaches have raised collars of precise diameters which act to ensure that the bronze inserts make firm contact with the walls of the original guide and that the diameters of the inserts are close to the final size required (photograph 9).

After broaching, Paul hones each guide in two steps. First, he uses the Sunnen Honall shown in photographs 10 and 11, which removes most of the metal needing removal

9

10

and provides a fairly good finish. After broaching and preliminary honing, the tops of the inserts must be trimmed off flush with the original guides (photograph 12).

11

12

Diamonds are Forever

For a perfect finish, Paul uses a diamond hone of the type shown in photographs 13 and 14. Once one of these hones has been through the guides, they really are perfect in terms of both size and finish. We keep mentioning size without telling you how it's

14

checked, which brings us to the matter of pilots.

A good head technician needs a large range of quality pilots (two Serdi machine pilots are shown in photograph 15). To size his guides accurately, Paul checks the

15

progress of his honing operations by periodically trying to insert a pilot of the exact size required. When it slides with just the correct amount of resistance, the guide is ready. Technically speaking, this method

A good head technician needs a large range of quality pilots.

13

of determining size is known as using a synchro-pilot as a plug gauge.

Inlet valves operate best with about .0015 clearance, with exhaust valves needing about .002 because of their greater expansion. Getting these clearances correct is important as too much clearance means accelerated stem wear and oil in the chambers, while too little clearance means that a valve could stick in its guide and cause all sorts of horrible problems.

There are two general types of pilot bits available: high speed steel and carbide. Although each can provide accurate sizing, high speed steel pilots are quite elastic (for metal) and can flex. On the other hand, carbide pilots are much stiffer and will break before any appreciable elastic deformation occurs.

The stiffness of pilots is important because if a pilot flexes the cutting operation for which it is a guide, it will become inaccurate. In preparing cylinder heads this means misaligned valve seats. Any flex in pilot bits is an even greater problem in the cutting of hardened inserts. These are so hard that an operator needs to apply fairly firm pressure to get the job done.

Compound cutters form all three angles of a valve job with the one bit. Combined with hardened seats, compound cutters virtually ensure that a high-speed steel pilot will flex. Carbide pilots are essential under these conditions.

Do You See the Light?

Once you've prepared the valve guides perfectly, their alignment with the valve seat cutter is essential. Each guide will have a slightly different alignment from its neighbour, so an operator can't just set one and assume that the others will be the same. Each needs to be carefully aligned using two bits of the same size.

Paul sets some white paper and a back-light behind the two bits and carefully adjusts the cutter until there is no light visible between the two. This is done in the forward and backward plane, and also from side to side for each valve seat (photograph 15). This ensures that every seat will be cut true (photograph 16).

Paul feels that the only way to gain almost perfect concentricity is to grind each seat (photograph 17). Each angle of the valve job

requires the appropriate stone (photograph 18) and, to ensure that these angles remain accurate, the stones must receive periodic dressing (photograph 19).

When it came to cutting the bad seat we'd asked for, Paul had some trouble producing it because he had prepared the guides so well. But with quite a lot of effort we

managed to get one with .007 run-out, way beyond the .0015 mentioned earlier.

Don't be Sucked In

We rang well-respected engine builder Peter Orger and told him about the article. He

agreed to let us photograph the head on his vacuum tester (photograph 20) because he also finds bad valve jobs annoying. Look at the needle in the photograph — right in the green. This seat is 0.007 out but still passes a vacuum test. This happens because of the

To ensure that these angles remain accurate, the stones must receive periodic dressing.

length of the valve stem. A guide can be well prepared with close tolerances but, because the valve head is so far away from the end of the guide, there can still be enough movement for the vacuum to pull the head across and make it seal.

As Peter pointed out, vacuum testers are good for establishing that a head has certain problems, which is why he has his mounted on the front counter. But as a means for checking valve seat concentricity, they are woefully inadequate. The only reliable means of checking concentricity is with a run-out gauge (photograph 21).

If you are about to pay for a valve job, the two most intelligent (and possibly annoying for them) questions you can ask a shop are, what final run-out do they guarantee to achieve, and do they check their work with a run-out gauge? The answers should help you make up your mind about whether to give them the work.

You can reach Eastern Cylinder Heads on (03) 9887 2560 and Orger Engines are on (03) 9729 9677. Thanks to both companies for helping with this article.

Camming it Up

One of the most important ingredients in any high-performance engine is the camshaft

Camshaft selection is possibly the most crucial factor in tailoring anengine's performance to your needs. The higher the horsepower you want from an engine, the more crucial is the camshaft design.

For years I have attended race and drag meetings, car shows and displays and have joined in countless conversations about particular camshafts, their grind, duration and lift. But what does it all mean?

For the purposes of this article we will pick a particular grind camshaft, say a Heatseeker cam, and work through its operation to see where all the facts and figures come from.

Note that the camshaft rotates at half the engine speed and all specifications are in crankshaft degrees.

The inlet opens

At 30 degrees before the piston has reached top dead centre (30/70 cam — means 30 for intake, 70 for exhaust) the inlet valve begins to open and allows the fresh air/fuel mixture to enter the cylinder in preparation for the power stroke.

The valve will not reach full lift until the piston has reached 110 degrees after top dead centre, the piston is at this stage approaching the end of the power stroke.

Grinding a billeted cam to specs.

INLET OPENS

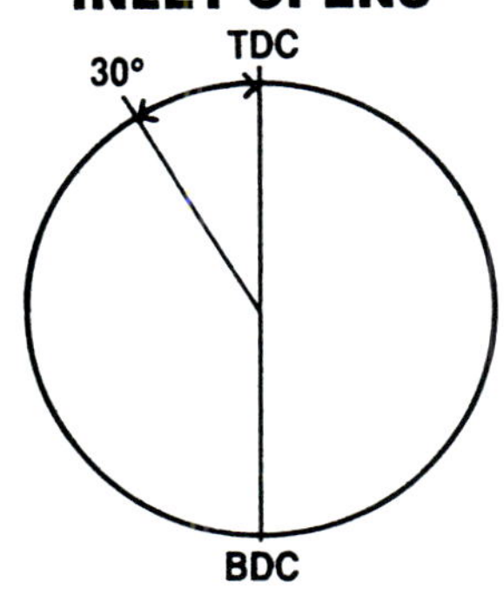

FIGURE 1

The exhaust

At this point the exhaust valve has nearly completed its run (it opened at 70 degrees before bottom dead centre).

EXHAUST CLOSES

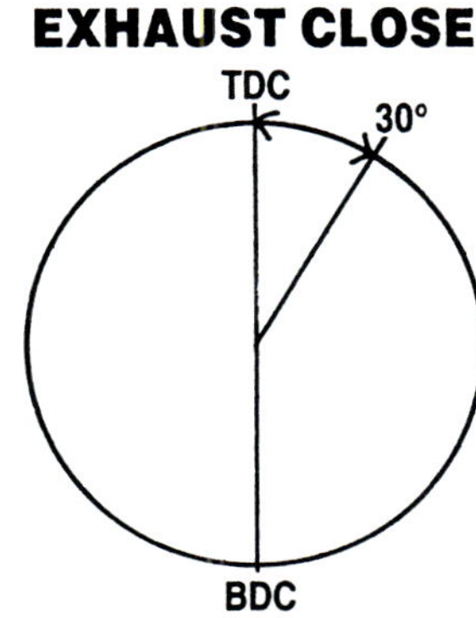

FIGURE 2

Fig 2 shows that the inlet and exhaust valve were open simultaneously for a period of 60 degrees — 30 degrees before top dead centre and 30 degrees after top dead centre. This situation is called valve overlap, therefore, this 30/70 cam has a 60 degree overlap.

The inlet closes

The inlet valve has been open from 30 degrees before top dead centre (BTDC) and will remain open until 70 degrees after bottom dead centre (ABDC). As the piston travels towards BDC it creates a vacuum and begins to suck a fresh mixture in. After it reaches BDC the piston begins to move up towards TDC. At 70 degrees ABDC the inlet valve fully closes.

INLET CLOSES

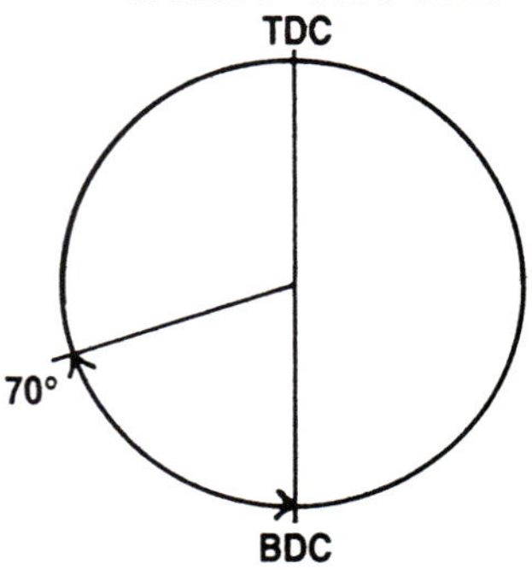

FIGURE 3

The exhaust opens

The piston is now approaching TDC and in doing so is expelling the leftover exhaust gases. The exhaust valve opened at 70 degrees BBDC and will not complete its run until it has passed through TDC and up to the 30 degrees ATDC mark.

When this happens the inlet (which has been open since 30 degrees BTDC) will have already been open for 60 degrees (the overlap period) by the time the exhaust fully closes and the whole cycle begins again.

EXHAUST OPENS

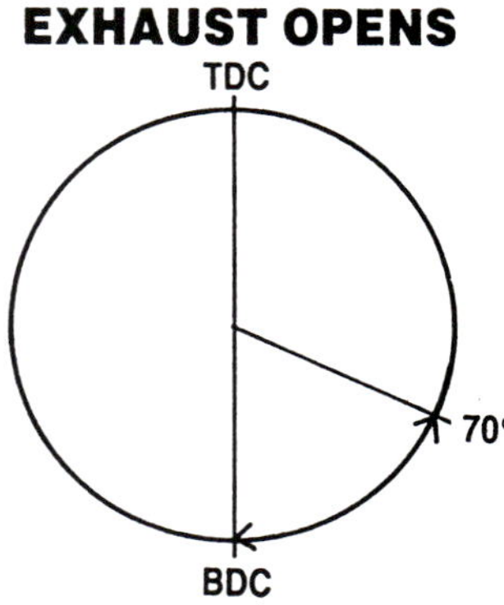

FIGURE 4

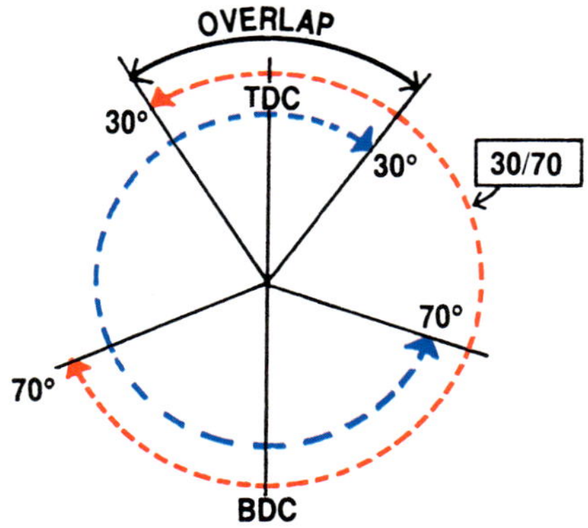

FIGURE 5

INLET CYCLE - - - - - - -
EXHAUST CYCLE — — —

Camshaft specification and jargon

For the purposes of this exercise we will be using a Heatseeker Pro-Series performance cam specifications card supplied by Watsons Camshafts, Boundary Rd, Peakhurst.

Watson Camshafts sells an extensive range of billeted and reground camshafts for both on and off street applications. It sells a tried and tested product.

Stage: This tells you that the camshaft is at a particular performance level. Remember that different camshaft designers have different ways of rating their camshafts, so stage 4 of XX brand may not be as wild as stage 3 of ZZ brand.

Lift: This is the actual lift of the cam, see Figure 6.

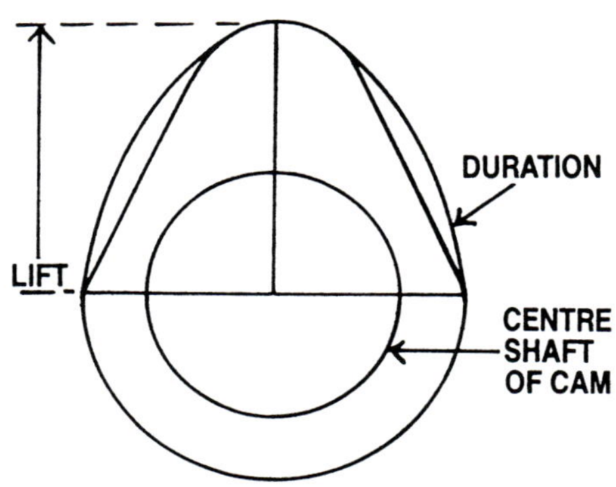

FIGURE 6

Valve lift: Valve lift is the cam lift multiplied by the rocker ratio minus the valve lash. The rocker ratio on a standard Holden V8 is about 1.5 (for every inch the pushrod lifts the valve will move 1.5 inches).

Valve lash: Engines with hydraulic lifters do not experience valve lash, this term is only used for engines with solid lifters. This figure is the tappet setting. In this case it is nil because it is a hydraulic camshaft.

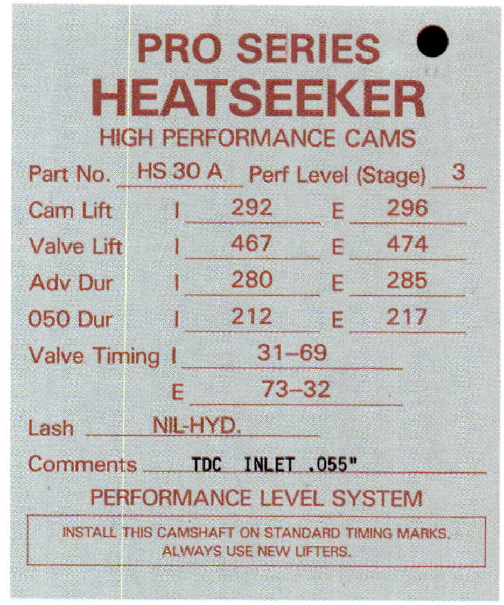

PRO SERIES
HEATSEEKER
HIGH PERFORMANCE CAMS

Part No. HS 30 A Perf Level (Stage) 3

Cam Lift	I	292	E	296
Valve Lift	I	467	E	474
Adv Dur	I	280	E	285
050 Dur	I	212	E	217

Valve Timing I 31–69
E 73–32

Lash NIL-HYD.

Comments TDC INLET .055"

PERFORMANCE LEVEL SYSTEM

INSTALL THIS CAMSHAFT ON STANDARD TIMING MARKS.
ALWAYS USE NEW LIFTERS.

However, with any solid or mechanical cam the valve lash is critical and must be carefully checked and rechecked when the engine is at normal operating temperature.

Advertised duration: This is calculated by adding the inlet opening and closing times plus 180 degrees. See valve timing card.

050 Duration: This figure is very useful for comparing different camshafts as it is measured at a common point, that being at 050 thou of lifter rise.

Valve timing: These are the points of inlet opening, exhaust closing, inlet closing and exhaust opening as explained earlier.

Comments: The comments section is reserved for any other information which may be necessary for a camshaft. This may, for instance, be information crucial to an engine builder who is degreeing a cam into an engine.

Roller camshafts

The fitment of a roller camshaft to a street car could be considered a waste of money to say the least as the full potential of a roller cam cannot be utilised on the street.

What a roller camshaft allows you that a normal hydraulic camshaft doesn't is the running of very high valve seat pressures, such as 300lb, at very high revs.

Roller cam Vs hydraulic cam.

Story/Photography: MARK ARBLASTER

Hydraulic cam lobes.

Performance street engines generally never exceed around 150lb of pressure at the valve seat. Roller camshafts have different lifters with round rollers which seat on the camshaft lobes. They relieve the high valve spring pressures exerted on the camshaft lobes.

The fitting of a roller camshaft with all the necessary extras can cost in the vicinity of $2000 dollars, money that can be better spent in other parts of the engine.

Roller cam lobes.

Application

By experimenting with various valve opening and closing times, camshaft designers are able to significantly increase horsepower potential, especially when utilising cylinder scavenging and gas inertia effects.

At high rpm the inlet valve can be left open for exceptionally long periods of time (85 degrees ATDC or more) in full-race-type designs. This has the effect of trapping more mixture in the cylinder and greatly increasing volumetric efficiency.

On the downside, although a high-duration camshaft might work well at 6000rpm, the same camshaft operating at 1600rpm could make the car undrivable — producing very poor horsepower and economy figures, not to mention a low engine vacuum, very rough idle and a considerable power lag under about 2500rpm.

Old lifters do not go with new cams. Note the cam severely worn lifter face.

When an old lifter is fitted to a new cam the edges of the lifter instantly gouge the cam lobe.

In general, standard engines only require longer-duration, low-lift camshafts, which allow the engine to breathe a little easier, to produce more power.

When replacing lifters ensure they are standard height or rocker collars may need to be machined or skimmed.

Fitting a short-duration high-lift cam will take torque away from the bottom end (low-rev range) and will not allow the engine to produce power until say 2700-3000rpm, and run up to 5500-6000rpm before peaking.

This would not be considered a practical powerband for a standard street car which would spend a lot of time in traffic and at low revs.

The important point to understand is that bigger is not always better. Such a camshaft would be better suited to a circuit car which spends all its time within a limited rev range.

Holden V8s are particularly harsh on high-lift camshafts (due to their poor oil circulating qualities) and will generally demolish lifters and cam lobes after a very short time, especially when fitted with high-pressure valve springs.

Points to remember

1. Always use new lifters when installing a new camshaft regardless of the age of the old ones. New lifters are slightly convex in design and need to bed in on new camshafts.

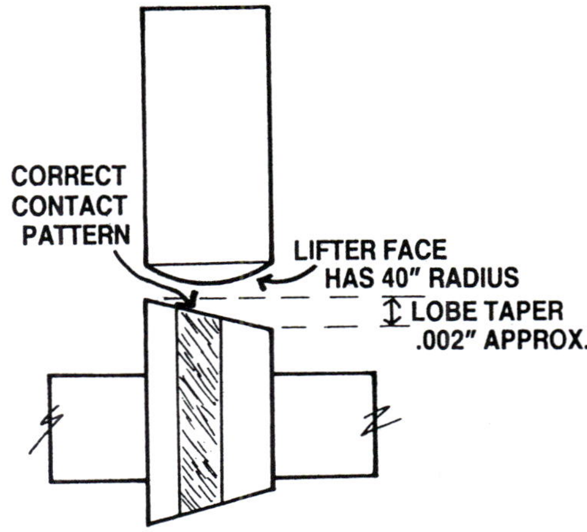

FIGURE 7

When they are bedding in, a certain amount of the face of the lifter will be worn away during the process. If old lifters are fitted to a new camshaft there is nothing left to wear on the old lifters and the new camshaft could be destroyed within a very short period.

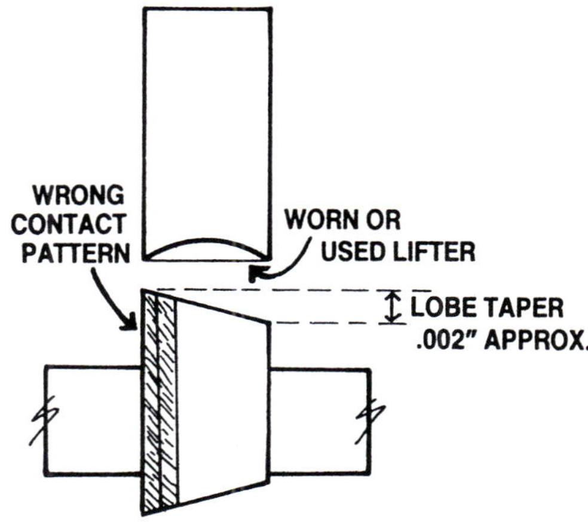

FIGURE 8

2. Having any machining performed on your heads or block can affect lifter pre-load. On hydraulic camshafts you must maintain a lifter pre-load of between 0.025-0.040 thou.

Engines with non-adjustable valve trains (Holden V8, Cleveland V8 etc) must have their rocker fulcrums shimmed or machined down if necessary to obtain the correct lifter pre-load. Running an engine without the correct pre-load can seriously damage the engine and result in poor performance.

3. When installing a camshaft in a Holden V8 do not line up the dots on the engine. Holden engines must have nine links from the markers on the crank and cam gear between them. Consult an expert or a factory manual on the correct alignment of Holden camshaft timing.

4. Consult an expert when buying a cam and don't exaggerate your needs and the work done to the engine as every camshaft has a very specific application and grind suited to a specific purpose.

E n g i n e

A V6 will often fit where once sat an anaemic four. This is a HiLux with a Chevy in it — but you get the general idea.

How weird and wonderful do you wanna get? Mazda 13B twin-turbo rotary in a BMW 320i!

Words & Pix by Godfrey Towns

Fitting a modern EFI engine into an older body makes a lot of sense. This is an FC Holden with a VL Commodore straight six. The swap was easy except the tranny tunnel needed enlarging.

Turbo'd Fiat 2-litre in a Datsun 1200 ute gives fair-dinkum 10-second quarters.

Putting a big-power engine into an ordinary body is the time-honoured way to achieve performance. You know what I'm getting at — V8 Toranas, EFI V8 early Falcons, Jap import turbo sixes and fours into pretty well anything. A heart transplant can revolutionise performance in an achievable (and affordable!) way.

It's all very well to see what others have done, but how about doing it yourself? We'll assume here that you're working on a rear-wheel-drive car, not front-wheel-drive (or all-wheel-drive) machine.

Before doing any DIY development work, make sure there's not an adaptor kit already available off the shelf. New engine mounts, bellhousing and clutch bits might be yours with the plonking down of some dollars, saving you a huge amount of time. Dellow Automotive and The Castlemaine Rod Shop are two sources to try for adaptor kits.

Making sure it'll fit

The first thing to do when considering an engine swap is to check that the engine will actually fit in the car. Obviously, if you're prepared to do major surgery, pretty well anything can be fitted into anything! But here we're assuming you still want to keep the inner guards, firewall and chassis rails pretty much the same as they looked when the factory made them.

There are three vital dimensions to measure: the distance between the chassis rails, the height of the engine and its length.

If the engine is wider than the space between the chassis rails you're basically stuffed! But make sure you actually do the measuring, though. Some cars are a lot roomier in this dimension than you'd first think — and some are the other way around. A Mazda RX-7, for example, has very wide chassis rails, even though the factory engine is physically quite small. Cars that came out with V6s and V8s (not necessarily for our market) are likely to have wider engine bays than those bodies available with only in-line fours and sixes.

The height of the engine is probably the least important of the three measurements. If the engine is too high, a bonnet scoop or bulge can often be used to cover the carb or injection plenum. Alternatives include lowered engine mounts. However, the length of the engine is important if it's fit easily. If it's a matter of 5-10cm

Swaps

DIY heart transplants

The numberplate says it all — and it's a fully legal, inspected swap. As you'd expect, the brakes have been substantially upgraded, too.

Australian Fairlane with Mustang EFI V8 is an easy engine swap that gives outstanding results.

US-import Ford 5-litre V8 isn't so much a swap as a bolt-in for the local Ford V8s.

or so (not 20 or 30), the radiator can be moved forward and thin thermo fans used.

The other area to measure is the transmission tunnel, but this is best done by trial-fitting the engine and tranny.

Test fitting

Because there are so many dimensions that need to line up, after you've concluded from some basic measurements that the engine's likely to fit it's then best to do a trial-fitting.

Unless the new engine bolts straight up to the tranny (pretty unusual but not unknown), it's best to slot both a new engine and transmission into place as one unit. To slide the whole lot in, it sometimes helps to first remove the crossmember (that's the hefty pressed-steel piece the engine mounts bolt to). With the engine and tranny in place (suspended by an engine hoist) you need to decide what you're going to line everything up from.

Let's say the transmission length is pretty close to original. That means the tailshaft is going to be darn near stock in length. You might decide then to get the powertrain into the position that allows you to use this stock tailshaft. Everything then gets worked out from there — the position of engine mounts, gearbox mounts and so forth. But if getting the tailshaft factory in length puts the engine halfway through the firewall you can scrap that idea!

The things you can line up first and then work from are:

* the engine mounts
* the position of the engine in the bay
* the gearbox mounts
* the position of the gearbox lever
* the tailshaft length.

Taking each of these as the starting point has advantages and disadvantages — but get at

Engine

You'll certainly need an engine hoist or block and tackle to do an engine conversion — the engine will need to be juggled around in the bay until its correct position is found.

A big V8 in a little body may mean there's no place for the exhausts! Here they exit through the inner guards — fine on a drag car but don't even think about it on a legal road car.

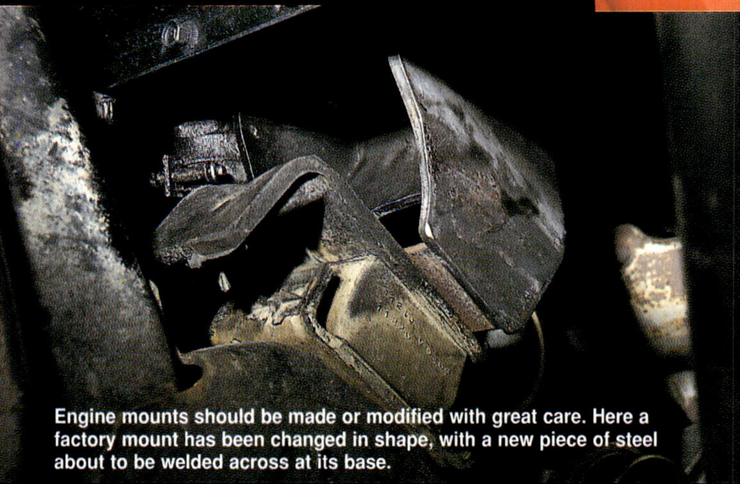

Engine mounts should be made or modified with great care. Here a factory mount has been changed in shape, with a new piece of steel about to be welded across at its base.

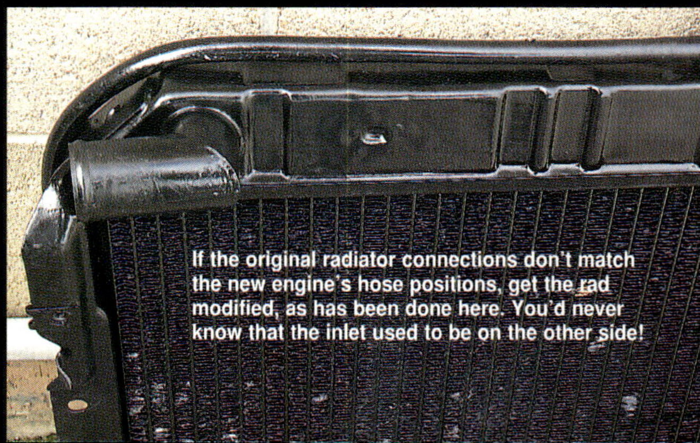

If the original radiator connections don't match the new engine's hose positions, get the rad modified, as has been done here. You'd never know that the inlet used to be on the other side!

A too-tall engine can be hidden under a bonnet bulge or scoop. Unlike width, the height of the engine isn't a vital measurement.

least one thing to line up or you'll have to adapt them all!

It can be helpful to put a block of wood on top of the (re-installed) crossmember and rest the engine temporarily on that. If you use a block of the right size, the engine can be located close to its final position, allowing the checking of all the clearances: bonnet to carb, firewall to engine, fan to rad and so on. From a handling point of view, the best position for the engine is as close to the firewall as practicable.

Engine and gearbox mounts

it's unlikely the new engine's mounts will bolt straight to the old crossmember. That means new mounts will have to be fabricated — and it's an area where some people do horrible things!

First inspect both the new and old engine mounts. Can the old mounts be used with the new engine? Can the rubber block from the new engine's mounts be used with the old engine's bracket? Check all the possible mixing and matching options using factory-produced mounts

A turbo exhaust is much easier — there's almost always room for a large single-dump pipe.

Here a long radiator hose is required and it's been formed from two right-angled bends and a length of tube. Moving the rad connection would have been a lot neater.

first. But if they won't fit, you need to make your own.

Let's say the new engine's mounts are further forward in the engine bay than the old crossmember's bolt holes. One approach is to simply use a piece of flat steel, bolting this to the crossmember mounts and then bolting the new engine's mounts to it. But this is not a good idea: the piece of flat steel will be subject to big bending loads whenever you floor the throttle! I once saw an adapted engine mount that had a staircase of pieces of flat steel welded, one on top of the other. The mount ended up being something like 15cm away from the original point! Failure was inevitable.

In this situation it's better to heat and bend the original steel so the new offset mounting position is held in place with straight (but angled) flat steel. In other words, there's a straight load-bearing path between the engine and crossmember. Some gusseting (triangular reinforcing pieces) welded into the corners would give this sort of bracket immense strength. If fabricating from scratch, use at least the same gauge material as in the new engine's mounts. At the gearbox end, the story's much the same. Either carefully adapt the factory mount or make a new one from heavy-gauge steel.

Gear lever position

If the tranny location will move the gear lever position, that's probably not a big deal — with some provisos. A few cms will make stuff-all difference: you probably won't even need to cut the hole any bigger in the tranny tunnel. However, if you're talking maybe 10cm, mods to the tranny tunnel and console will be needed. You might also want to sit in the car and pretend to drive with the gear lever this far removed from standard. If it's always gonna be a pain in the butt (or arm), maybe putting it there isn't a good idea.

If the tranny tunnel does need to be cut, a metal-cutting blade in an electric jigsaw will give the neatest result. Don't forget to file away the sharp edges, and then paint the raw steel with a rust preventative. Making the console then fit back in place can be difficult — one easy way is to fit a larger gear lever boot that covers both the original and the extended opening.

Exhaust

Generally, custom-made extractors will be needed to fit the new contours of the engine/body combination. On turbo cars, a single dump pipe out of the turbo is easy to fit.

Radiator and Fan

The new engine and the old radiator probably don't have a good relationship going when it comes to coupling — the water hoses, that is. If the rad has the outlet (or inlet) on the wrong side, have a radiator shop change it over — a heap neater than running long cross-over hoses. When sourcing hoses, bend up a guide of what's needed from welding wire and then go to a big auto parts shop that will let you browse through their selection of hoses. A moulded hose that fits is so much better than one of those 'universal' convoluted hoses that scream "bodgie engine swap!" whenever you open the bonnet...

Most factory cars have an engine-driven fan up close to the radiator, all surrounded by a shroud. If the new engine has a viscous-coupled fan it makes sense to leave it in place rather than fit new fans. However, it's likely the old shroud won't fit any more. Searching through a wrecking yard or modifying the old one are two options.

Legality

We haven't covered the legal implications of an engine swap here — but you'd better. Ring your local registration authority to find out what needs to be done and what is allowed — almost certainly a brake upgrade will be a requirement. However, if your workmanship is up to scratch and the end result looks factory-neat you're unlikely to have many legal problems in a well-braked car featuring a non-standard underbonnet mill.

TORQUE ABOUT POWER part 2

CHASSIS DYNOS

By Paul Tuzson
Dynotuning

Last issue we looked at engine dynos, so this time we give chassis dynos a going-over.

Like the engine dynos we discussed last time, chassis dynos are a means of artificially creating a load to simulate operating conditions. Again, a retarder is used to create a drag that opposes driving motion. As with engine dynos, a number of different types of retarders have been used to accomplish this task. The two main types are hydraulically retarded and electrically retarded. Whereas hydraulically retarded units seem more common on engine dynos, the opposite is true for chassis dynos; most of the units you'll encounter are electrically retarded.

An electric retarder operates by using a strong composite magnetic field. A number of field-generating coils are arranged in circular fashion and these create drag on a couple of large iron disks attached to the rollers in which the driving wheels of the vehicle being tested are located. As the amount of current supplied to the coils is varied, the strength of the magnetic field is correspondingly increased or decreased. This adjusts the load opposing the driving motion of the wheels.

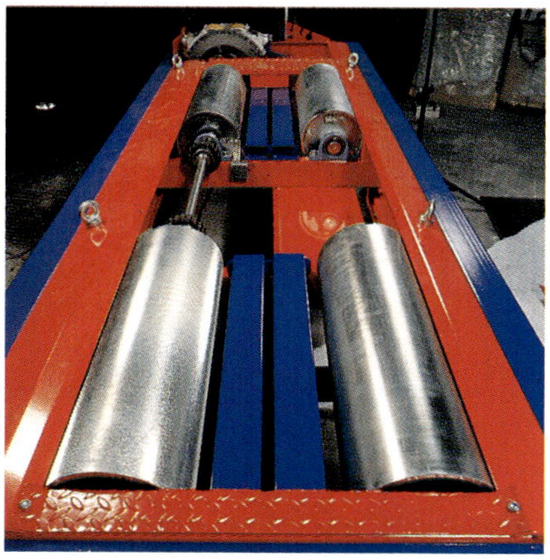

The arrangement of rollers and retarder on a Dyno Dynamics machine.

IT'S A DRAG

As the iron disks rotate, they try to drag the coil section of the retarder around with them. The coil section has an arm attached to it and is free to rotate very slightly. The end of the arm is located on a load cell and, the harder the iron disks try to drag the coils around with them, the harder the end of the arm presses down on the load cell.

The load cell is a block of steel or aluminium which contains electrical elements that conduct electricity in varying amounts, depending on the strain applied. The length of the arm and the pressure on the load cell give a reading of the torque the drive wheels are creating. Some manufacturers quote

The heart of any dynamometer is the retarder. This is an electric, or eddy-current, retarder. The iron rotors can become very hot during extensive testing. Like a disc brake rotor, cast-in vanes circulate cooling air to dissipate this waste energy.

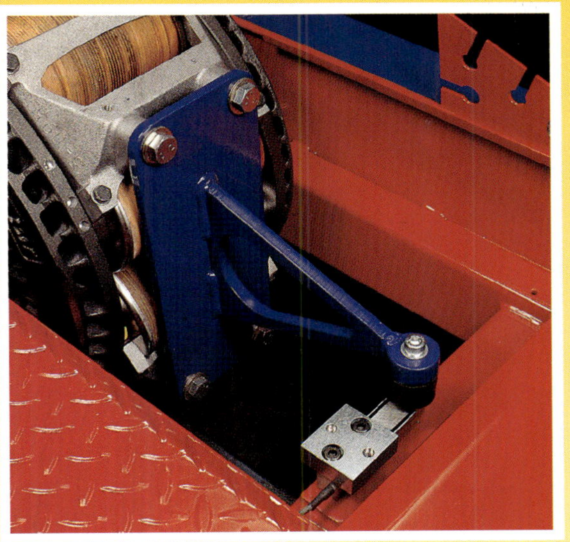

This shows how the torque arm acts on the load-sensing cell.

figures for both torque and power while others make a further calculation that converts the basic torque reading to something called tractive effort. This is a measure of pulling force generated by the vehicle. Dyno Dynamics says it's best to think of it as connecting a giant spring balance between the vehicle and a wall and taking force readings from that.

Measuring speed is fundamental to calculating power and general performance characteristics. Modern dynos use a Hall effect sensor to take signals from a spiked disk just like the crank sensor on a performance car. This must be tucked out of the way somewhere to avoid damage. The rollers must have a friction surface on them to provide a secure grip for the tyres, otherwise slippage will occur. The best way of creating such a surface is to cut it into the rollers, that way it can't wear off.

If the tyres do slip, the performance figures will be inaccurate and the tyres will suffer more wear than they should. Slipping tyres can also cause the vehicle to surge in the rollers. This can be dangerous and can give the impression there's something wrong with the fuel supply, which can be time wasting.

DIFFERENT TESTS

There are different ways the load can be applied to a vehicle. They're similar to those we described in the last issue but we'll run through them again for new readers.

On a chassis dyno, the different testing modes are speed, load and ramp. Setting the dyno to speed mode means the retarder will automatically increase or decrease its resistance to maintain the selected road (roller) speed no matter what the operator does to the throttle. This is the most commonly used testing mode.

Testing in load mode means the retarder will adjust itself so the load on the rollers is always the same no matter what speed the operator chooses. In load mode, the speed of the vehicle will rise and fall according to the throttle setting. Finally, there is ramp mode in which the speed of the vehicle is controlled automatically according to preset parameters.

A ramp test is similar to the step test described in the previous issue.

However, the dyno doesn't pause at the pre-selected reading points. The acceleration between the start and end of the test speed range is continuous and the computer quickly takes a reading as each test speed is reached. This method results in low readings because of inertia. Inertia is resistance to change. Constantly accelerating all the moving parts without pausing uses some of the torque produced to overcome inertia and results in reduced readings.

Fortunately, inertia also works in reverse, so the way to get accurate readings with a ramp test is to continue taking readings at the selected rev points as the car slows down. Under these conditions, the inertia of the moving parts makes them want to keep moving and this tendency is read by the load cell as extra torque which creates an artificially high reading. If the low readings taken during acceleration and the high readings taken during deceleration are averaged, the result will be the true power for the vehicle.

All these different methods of testing offer an operator the ability to test virtually any aspect of a vehicle's operating characteristics. Previously, this could still be done but it took much longer because frequent test driving was necessary. From the operator's point of view, having a dyno means he doesn't have to risk driving customers' cars around performing road tests. More importantly, getting your car tuned by someone with a dyno means ►

Accurate speed readings are essential and trigger wheel should be mounted where it is least likely to be damaged.

The rollers themselves are important. Dyno Dynamics insists that the grip pattern should cut into the surface of the roller for best results. They have their own unique style.

that no one has to take your pride and joy out and mix it with the motorists, which is a definite plus considering how much time, money and effort you've probably spent getting it on the road.

OVERSTRESSING

What about stories of overstressing an engine on a dyno? Dyno Dynamics describe it this way: driving a car along the road at a particular throttle setting results in a certain speed. If the car comes to a hill and the driver doesn't open the throttle further, the car slows down. The steeper the hill becomes, the slower the car goes. It's the same on a dynamometer. Increasing the load against which the wheels must work is the same as encountering a hill of increasing gradient. The car won't be stressed any more on the dyno than it would be on the hill, as the operator opens the throttle further to compensate for the decreasing speed.

Rollers must be perfectly balanced and run in properly mounted bearings so the dyno doesn't contribute any vibrations.

Chassis dynos are useful for all sorts of tests. The fact that an operator can stand and watch what's going on under the bonnet while the car is effectively driving down the road at 100km/h or more is a definite advantage. I've seen guys wondering why their cars lacked power and eventually discovering the secondaries weren't opening. They could have found out about this much sooner by getting a dyno tune. This should be the first step whenever you buy a car or after you've made any modifications. Most dyno tunes are relatively inexpensive

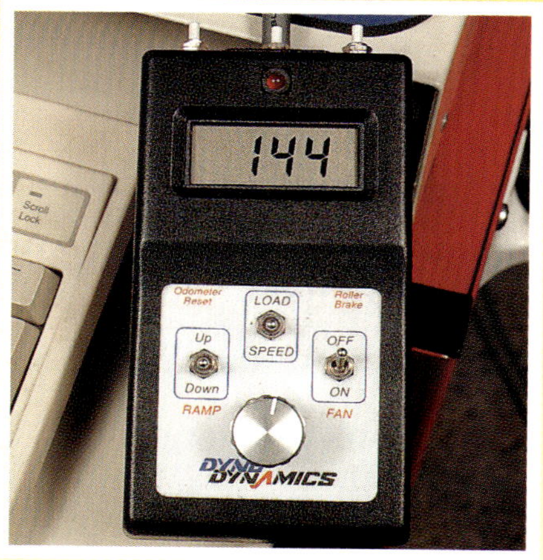

A hand-held control unit. See the main text for an explanation of the features.

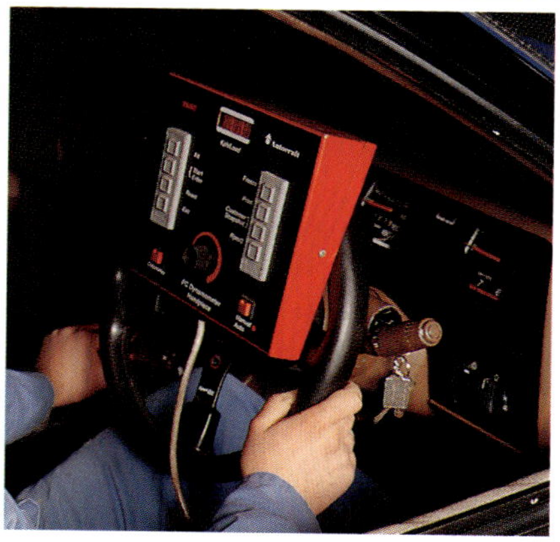

because of the short amount of time required to find any faults.

The idea of testing for comparisons after you've made modifications introduces the issue of accuracy. Will the results be valid for direct comparison of before-and-after figures, particularly if there's quite a bit of time between the tests? The answer should be yes. You'll probably settle on someone you

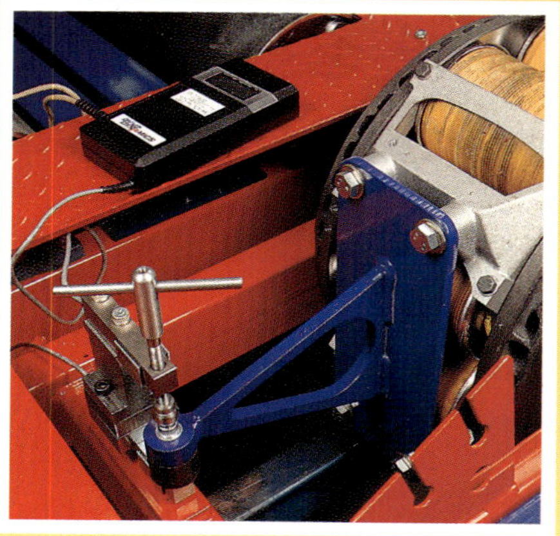

Calibration equipment that ensures all dynos will read the same. Different manufacturers have different systems.

like and constantly take your car to the one place for all its evaluations. This will certainly eliminate variations from dyno to dyno.

HOW ACCURATE?

Accuracy on any one dyno is easily achieved. It's simply a matter of putting a pre-determined load on the load cell and adjusting the numbers to whatever they should be. In the past this was done by hanging a certain number of weights on a stand attached to the cell. This is a perfectly acceptable method and many dynos are still calibrated this way. However, when we went out to Dyno Dynamics they showed us their latest tool for doing the job.

It's a second load cell that's bolted to the one on the machine. A screw is turned down until a specific load is indicated on a hand unit, at which point the dyno is adjusted to read as it should. Correct speed indications are also vital, so the output of the speed indicator is checked against a number of known frequencies generated by a hand-held unit.

Correction factors are another area of concern. Of all factors for which compensation should be applied, inlet air temperature probably accounts for 80 per cent. Testing on hot or cold days can mean up to 20 per cent difference for the power rating of the same vehicle. As Dyno Dynamics points out, that's more than the power difference between a 253 and a 308. Eighty per cent of the remaining 20 per cent needing correction is probably taken car of by adjusting for relative humidity. This affects the ability of air to hold fuel in suspension. The more moisture there is in the air, the less fuel it can hold.

This is an eddy-current retarder for an engine dyno designed for use with high-performance racing V8s. Because the output of these engines is so great, air cooling of the retarders won't remove the heat quickly enough. In this case, water cooling is needed. The iron rotors run in the chambers sealed by the red bands.

TORQUE ABOUT POWER part 2

CHASSIS DYNOS

These factors should be compensated for automatically and the values for these or any other correction factors should be noted on printouts from tests so that differing settings can be compared.

Air/fuel ratio meters are used in conjunction with dynos, so finding fuel supply problems is fairly easy. If your car is lean according to the meter, a pressure gauge at the carb inlet that reads normally tells you there's no problem up to that point and indicates that the carburetor settings must be wrong.

All these things can be fitted to a car which is then road-tested but it's far more difficult and risky for your car. All manner of fuel system testing can be carried out on a dyno.

IGNITION MATTERS

Finding the right ignition timing can be tricky and risky because of the speed at which detonation can cause damage to a highly loaded motor. On a dyno, you can adjust the distributor under the bonnet with the car running at the load and speed settings it will see in service. Timing can be backed off a bit at the first hint of detonation. Plotting the power curve on a graph clearly shows any areas of poor performance and timing, and other factors can be adjusted to create a perfectly smooth curve throughout the range.

Besides the more obvious sorts of checks and adjustments we've been mentioning, there are other ways to use a chassis dyno. For instance, finding vibrations can be much easier. Suppose you have a vibration due to an unbalanced wheel. If, when you run the car on the dyno, the vibration disappears, you know it's in the non-driving wheels. If it's still there, you can swap wheels from front to rear fairly quickly until you find which one's the problem.

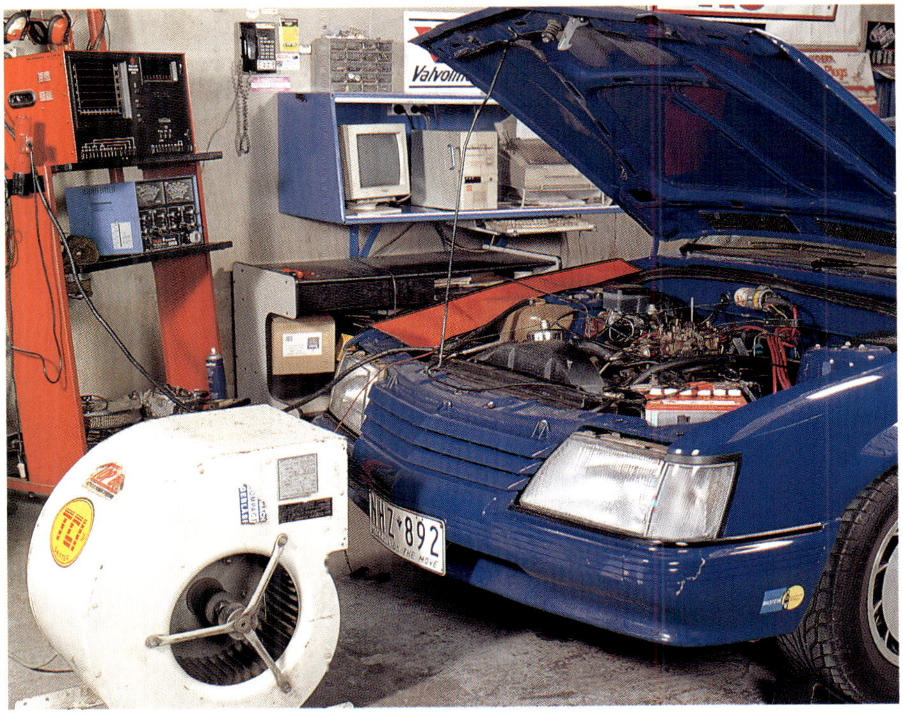

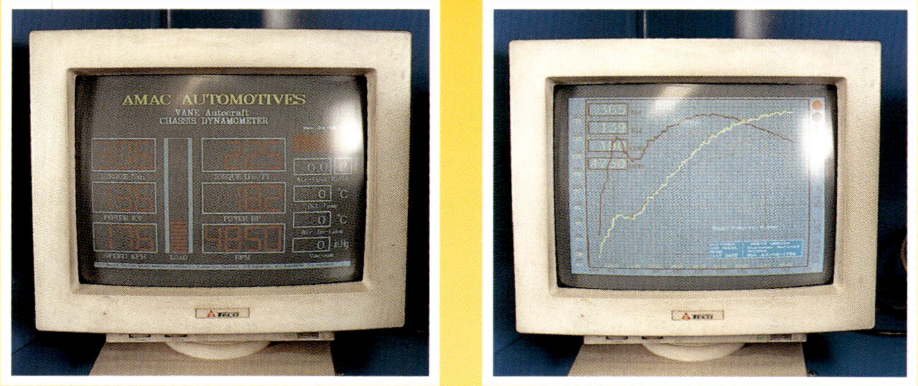

In days gone by, analog instruments used to tell the story. These days, everything is computer-controlled and displayed.

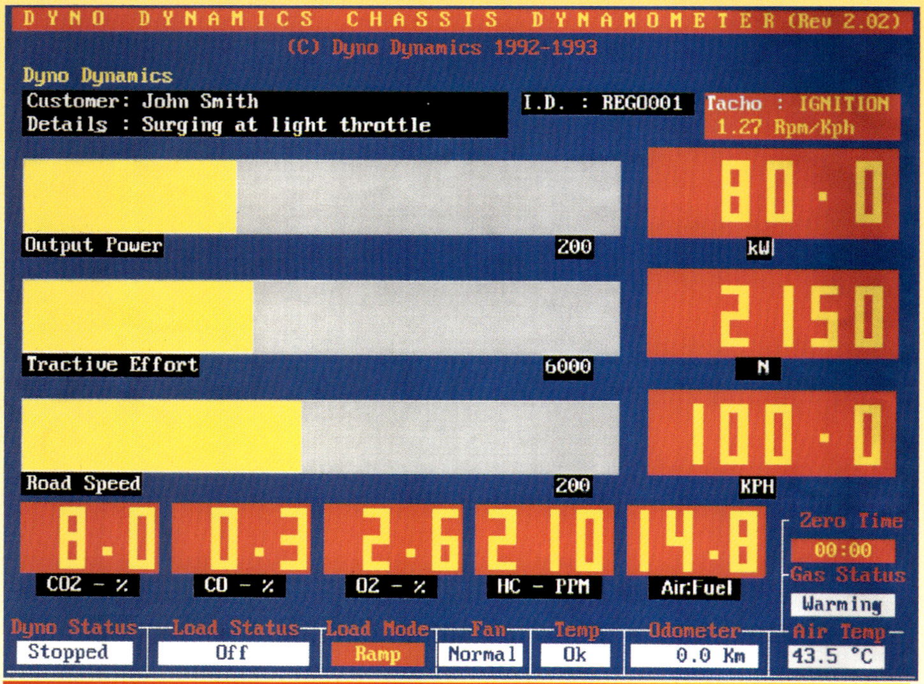

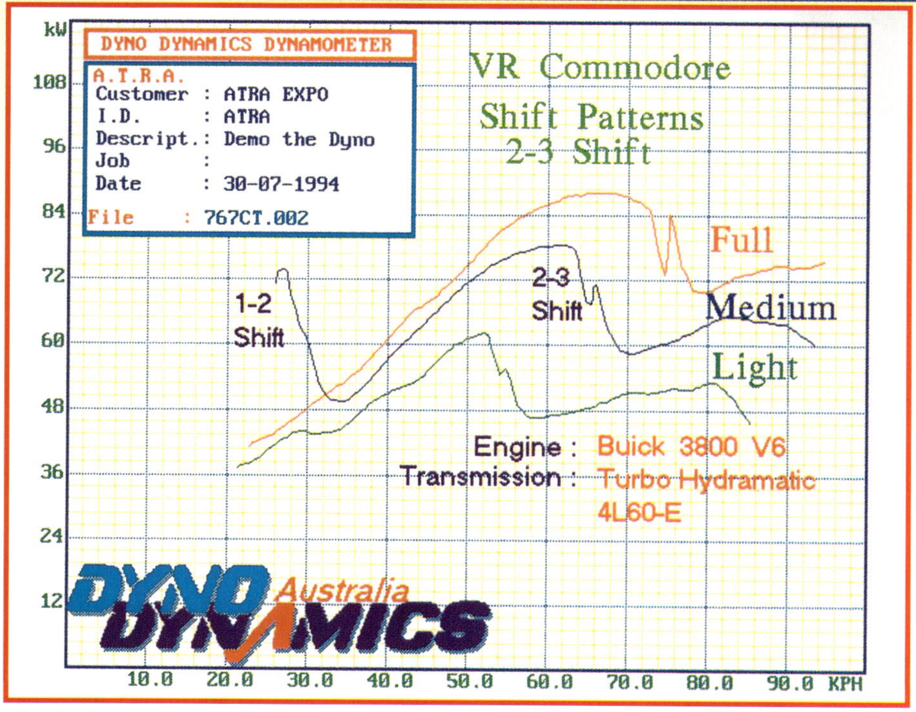

Graphs that overlay results from different tests are an excellent way of visualising the various aspect of your car's performance.

TORQUE ABOUT POWER part 2

CHASSIS DYNOS

I'M GETTING THOSE BAD VIBRATIONS

If you have a general vibration and you're not sure where it's coming from, the dyno can help isolate it. Suppose it occurs at a particular road speed. The dyno would be set to hold the problem speed and then checking could begin. If the vibration disappears, it's in a part of the vehicle that's stationary. If it's still there, the operator could vary the engine speed by changing gears and keeping the road speed the same. If the problem goes away, it's probably in the engine; if it doesn't, it's probably in the drivetrain sommewhere.

Noises can be isolated in similar ways. Dyno Dynamics says for this sort of checking to be possible the dynamometer itself must be free of vibration. This means all the rotating components must perfectly balanced and running in quality, well-mounted bearings.

Other things that can be checked include fan clutches, drivebelt slippage, combustion-chamber-to-water-jacket leaks, shift points for automatic transmissions and, of course, all the more commonly considered uses of a dyno like plotting power curves.

All things considered, the chassis dynamometer is one of the most useful tools available for sorting out your car when it's finally running. As soon as your project is finished, book it in for this finishing touch to make sure everything is right before you hit the streets.

Dyno Dynamics and Ozy-Dyn provided information for this article. We took the component shots at Dyno Dynamics and dyno shots at Adrenalin Performance (03) 9761 0355 and Amac Automotives (03) 9796 2444.

Street TECH

Do-it-Yourself Fuel Systems

by Andrew George

Fuel systems can look impressive like this small block Chevy in Geoff Garlan's '57 Chevy. It also features direct port nitrous oxide injection.

Car engines run using heat from the combustion of a fuel charge. To increase the performance of any engine more heat is required in each power stroke. Whether this is done by fitting blowers, turbochargers, larger carbies, extractors or head modifications, the basic aim is always the same — to burn more fuel. Since an engine runs best on a mixture of approximately 15 (actually 14.7) parts air to one part fuel, any increase in the mass of fuel burnt must not disrupt this ratio.

Getting more air into the engine is easy, as gravity will 'push' air into the inlet as the piston moves down on its induction stroke. But petrol is a different story. It has to be carried on the car, pumped from the tank, filtered, metered, then mixed with air. The art in making any car go fast is in moving as much fuel from the tank to the engine as possible. A major concern in this is design of the fuel system.

The movement of fuel towards the engine starts at the fuel tank. Tanks can come in all shapes and sizes. On production cars they are often placed to allow as much boot space as possible. On race cars the fuel tank is a vital component in the vehicle's performance. For example, the 'dropped' tanks seen on racing sedans play an important part in strategies of race teams. By delaying the need for the car to pit for fuel, petrol-only stops can be avoided.

On a drag car the task is slightly different, as volume is often not as critical but the object is still to flow as much fuel down the line as the engine requires. The relatively high accelerations achieved in drag racing make fuel pick-up an important issue. So at least one of the pick-ups is always towards the back of the tank. If you are only using one pick-up don't put it at the front. As the fuel is forced to the rear, it may suck air.

As you can see in the diagrams, there are many combinations of components that can and may need to be used. It all depends on the application. If using nitrous oxide (which is not a fuel) an extra supply of petrol is required. Often a separate line from the fuel tank is run for the nitrous oxide system alone. Basically, the more power the engine can make, the more fuel it will require.

To force the fuel from the tank towards the engine, a pump or a system of pumps will be required. Generally there are two types of pump: mechanical and electric. For racing, electric pumps are often preferred due to the increased pressure volume that can be obtained. All electric fuel pumps have to be regulated due to the increased output pressures. Their fitting is also more flexible since they don't need a source of mechanical effort to drive them.

However, if your street car already has a mechanical pump fitted, by simply enlarging the fuel line and fitting a low-pressure electric pump on the tank side of the mechanical

SINGLE PICKUP AT TANK — Recommended minimum line sizes.

The simple and most basic set-up for street machines involves using a single pump and filter to supply a four-barrel carburetor. Add-on nitrous system is shown in 'dotted line'.

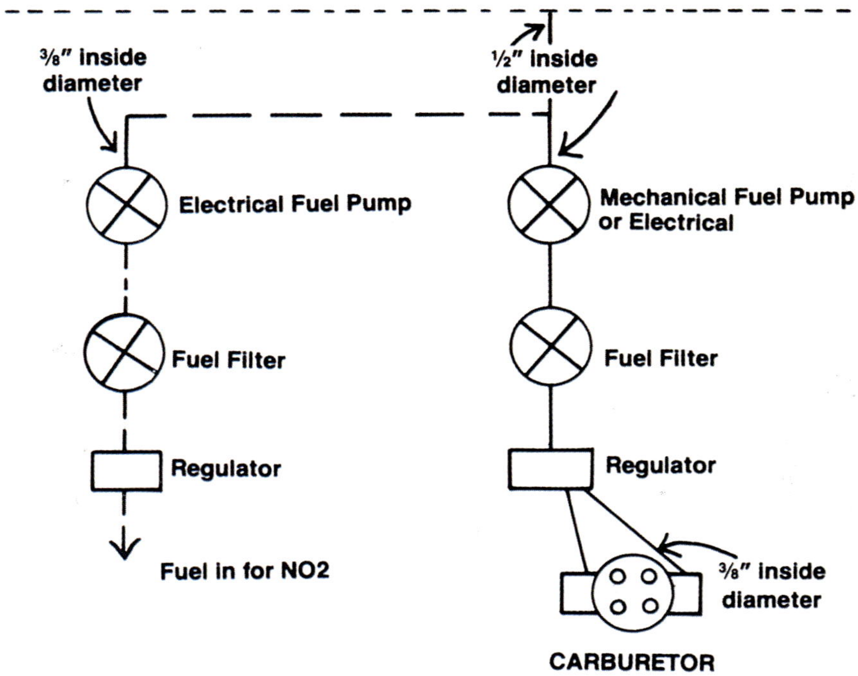

Note: Fuel filters and pump sequence can be reversed depending on manufacturer's recommendations.

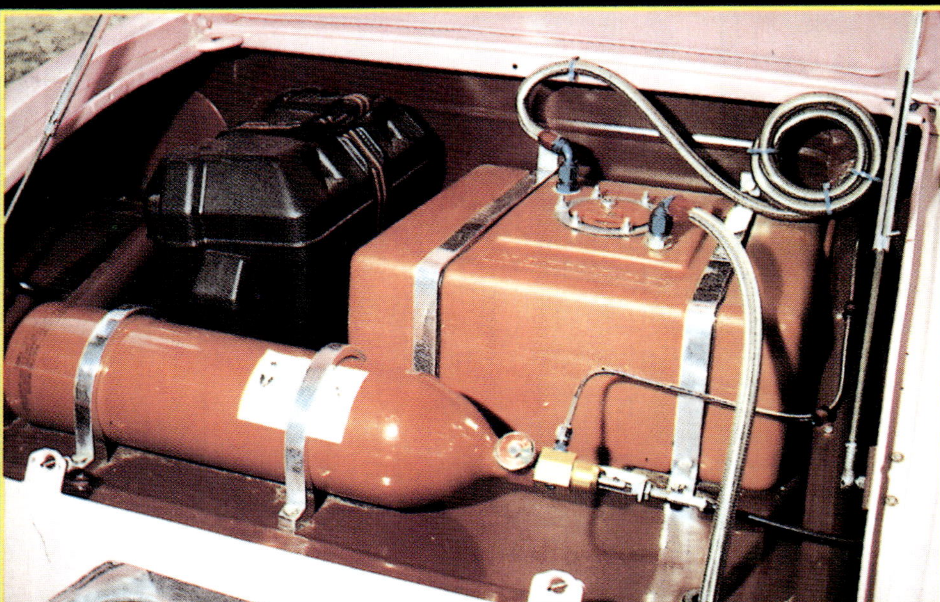

The boot area of Les Mathieson's competition-style Ford roadster is well laid out with its fuel cell, sealed battery box and on-board fire extinguisher system.

Source: Haddad Race Cars and Engines
16A Amberly Cres Dandenong Phone (03) 794 6599

pump, increased fuel flow could be obtained at a reasonable cost (see the relevant diagram). For most street applications fuel pressure should be at least 5psi throughout the full range of engine speed. Most mechanical pumps fitted to V8 engines should be able to handle this.

For racing applications, fuel pumps can be set up in parallel to obtain higher flow rate. The pumps used for this are nearly always electric, since a system can easily be designed to suit the engine requirements relatively easily.

However you decide to configure the system of pumps, the correct fuel pressure must be maintained over the engine's range of speeds. This is performed by a regulator. The regulator is normally positioned as close to the carburetor fuel inlet as possible. On multiple four-barrel carburetors several regulators may be required, depending on how critical correct fuel pressure is. For example, out cruising on a Saturday arvo fuel pressure would not be an issue as it would in the last half of a 200mph quarter-mile run.

As a general rule, most Holley carbs have a maximum needle and seat pressure of about 7psi, whereas the Rochester four-barrel can be regulated about 0.5psi higher.

The next important part of the fuel system has the function of keeping the whole system running as it was designed: the filter. By positioning the filter before the pump, water and dirt can be kept out of the pump. But this has its own problems as fuel must be sucked through the extra resistance of the filter. This often outweighs the benefits of keeping water out of the pump and usually results in filters being placed between the fuel pump and the carby. Wherever you put the filter, keep it out of the way of any damage. Just remember — if the fuel filter is upstream of the pump and is somehow damaged, petrol will be pumped out until the engine is shut down. When petrol comes in contact with hot surfaces it usually catches fire.

Next we have the mounting of the lines and all of the components. They must be mounted securely and suspended so the lines or components don't move in normal

This ex-Bathurst A9X Torana has a large fuel tank which holds 140 litres; note the low restriction filler neck.

DUAL FUEL PICKUP AT TANK — Recommended Minimum Line Sizes

This dual pickup can be used to supply a single four-barrel carburetor or two four-barrel carburetors. Twin carburetor set-ups actually require *less* volume than single carburetors. This is because twin four-barrel carburetor systems have twice as many fuel outlets to supply the same amount of fuel to meet the same power output.

Nitrous system can be added as shown in other diagram.

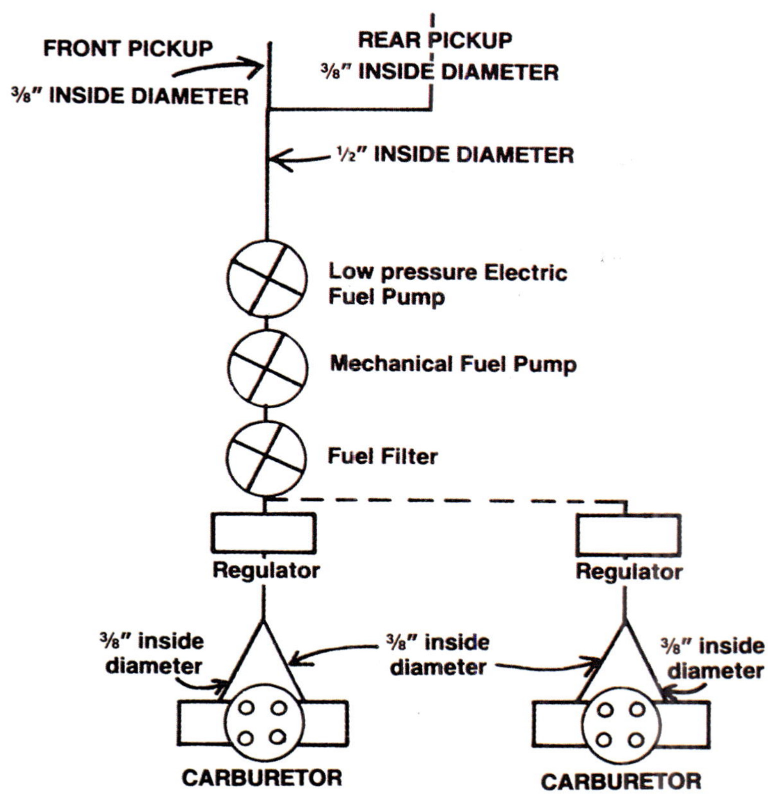

Note: Fuel filters and pump sequences can be reversed depending on manufacturer's recommendation.

A shot of the A9X Torana's fuel system using twin downdraft Weber carbies using an air box under the bonnet scoop. Note the two fuel regulators, one for each carby.

Fuel pressure gauges should always be mounted on the outside of the car for occupant safety.

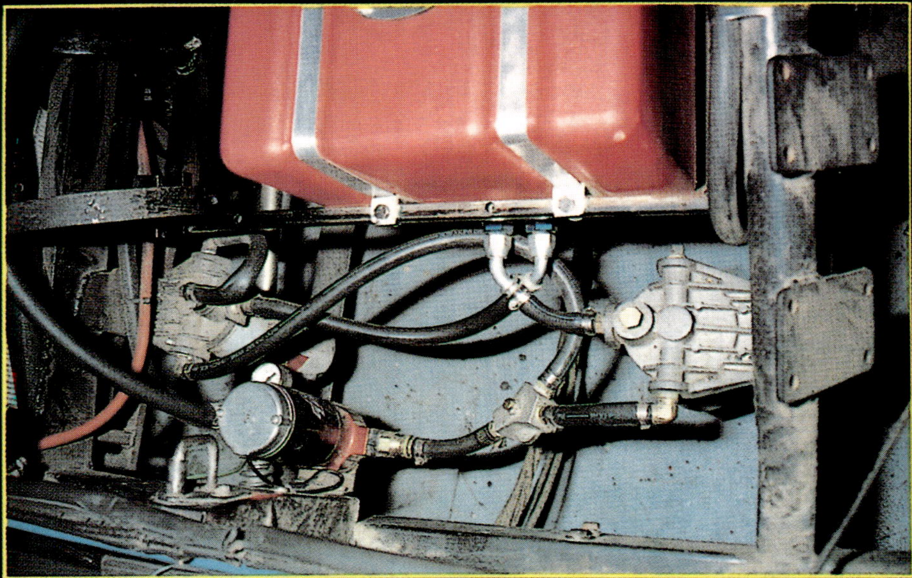

George Haddad's Capri features a fuel cell (red) with twin outlets at the rear, two filters and a Barry Grant fuel pump.

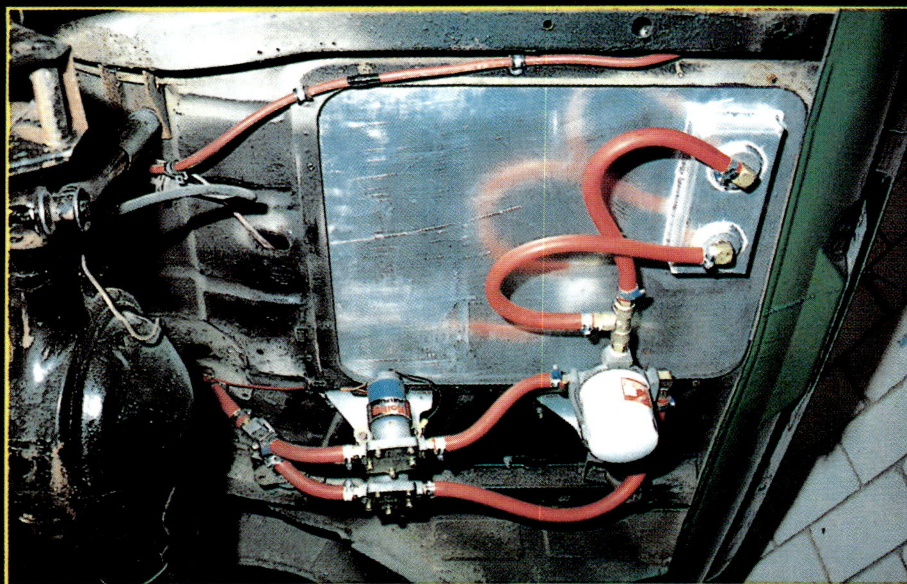

This full system from a competition Torana has two outlets into the one line until one of them is used for a future NOS fuel supply. Then through the filter and onto the dual pumps.

use. Fuel lines should be supported at least every 200mm around the engine bay area. Under no circumstances should fuel lines be routed through the cabin, not even for a fuel pressure gauge.

Finally, the other important aspect of the fuel system is in the design of the lines. By this I mean the routing and sizing of the lines. The tip in designing systems is to read the manufacturer's data. The volume and pressure rating of the components will be printed on the data sheets. Often manufacturers recommend certain systems for different applications. This is valuable information.

Fuel line sizing plays an important role in the design of a fuel system. Increasing the line diameter will increase the potential volume (and decrease wall friction) but will require a larger pump to keep up the pressure. Conversely, fitting a smaller line will increase the pressure but decrease volume.

Components such as filters have the effect of causing a drop in line pressure. To give some idea of what components can be used, diagrams of different systems have been included in the boxes. By looking at these configurations and reading the manufacturer's data, a ballpark set-up for your car shouldn't be too difficult.

POWER AID

So you've upped the comp ratio for more grunt and economy. But how do you stop that detonation?

BY GODFREY TOWNS

Back in the '60s, fuel — especially in the US — must've been a bloody different fluid. How else can you explain a factory Corvette with a 427ci engine and 425bhp from an 11:1 compression ratio? *11:1!* And then there was the Shelby Mustang with just a little 289ci V8 and 10.5:1 comp ratio ... and 306bhp! These days, the five-litre Falcon has just 9:1, while even the blueprint Commodore GTS-R has a measly 8.5:1 compression. So what's the big deal?

COMPRESSION RATIOS

We all know what a five-litre engine is ... don't we? A five-litre V8 has a cylinder volume of 625cc — that's 5000cc divided by its eight cylinders. So each cylinder has a volume of 625cc when the piston is at the bottom of its travel. There's not much room left when it's compressed the air/fuel mixture on the upward stroke; in fact, with a 10:1 compression there's only 62.5cc left when the piston's back up the top. With a 9:1 comp ratio there'd be a volume of 69.4cc left. So does it matter much? Well, yes it does. The higher the compression ratio the more energy gained from the fuel being burnt. And that means improvements in both power *and* fuel consumption. How much of a gain you get depends on your starting point. If you've got a comp ratio of 8:1, going up to 10:1 will give you a lift in power of 11 per cent, and remember, economy will also actually improve If you went up as far as the 11:1 of that '60s Corvette, you'd expect to improve your power output by nearly 17 per cent...

And increasing the comp ratio is as easy as shaving a bit of metal off the heads. You wanna high-comp ratio — then take about 40 thou off when the heads are being done!

2. While in the '60s big Yank iron ran factory comp ratios as high as 11:1, these days the five-litre Falcon has just 9:1...

DETONATION

OK, OK, now you're thinking, "So if I go up in comp ratio to 22:1 I'll get..." No, sorry, all you'll get is a blown engine! The reason makers are conservative these days with their compression ratios is that today's fuels have lower octane ratios than fuels of the past. While there are two different ways of measuring the octane of petrol (RON and MON), both are basically on about how easily the petrol gives rise to detonation. So what's detonation?

1. The higher you can make the comp ratio the more grunt you'll get from the engine. However, over-high comp ratios will also mean the engine will be prone to detonation ... and detonation kills engines.

Even though you may well think the air/fuel mixture explodes into action when the spark fires, what happens is actually more like a sped-up grassfire. If you set fire to the edge of a field of tall dry grass on a still day, the flame front spreads out from

3. ...while the V8 Commodore has a measly 8.5:1 compression. And that's with special knock sensors listening-out for detonation!

the point you applied the match to. One grass stalk sets fire to the next grass stalk and so on. The fire doesn't jump ahead of the flame front at all.

Well, that's kinda like the normal combustion process in a cylinder. The flame front gradually moves across the combustion chamber and, as it does so, the pressure of the burning gas increases. It's this gradual pressure increase that pushes down the piston, giving you the grunt to spin your back wheels!

But when detonation occurs, the flame front doesn't move this way at all. Instead, combustion occurs in front of the moving flame. It's like you've plonked down a big wad of cotton wool soaked in petrol. Even when the flame front is yet to reach it, it still bursts into flame. In a combustion chamber this sets up another pressure wave, which collides with the first one started by the sparkplug. The result is like a hammer blow to the top of the piston.

You can hear that hammer blow by the 'tink, tink' noise, but for you to hear it happening at all it means the piston and head are being violently bashed. If left unchecked, detonation can destroy the pistons, crack the heads and burn the valves. Not nice. That's why — even with special sensors listening for detonation — manufacturers now specify low-comp ratios on their production engines.

5. Bolt on a blower like on this 6.6-litre big block...

4. Running nitrous will make the engine more prone to detonation, although the higher cylinder pressures are counteracted in part by the cooling effect of the nitrous oxide.

CONTROLLING DETONATION

Detonation occurs because the combustion pressures get too high. That's also why if you fit a blower you're much more likely to get detonation. With more air/fuel mix forced into the cylinders the combustion pressures get even higher. Over-high combustion pressures cause the fuel/air mixture to explode into action ... read the bit above again if you've forgotten it! Detonation is also more likely if the combustion process gets too hot, just like with the grassfire.

6. ...or on this 3.8-litre V6, and the chances of the engine detonating are much higher.

7. Detonation can be controlled by retarding the time at which the spark fires. This reduces combustion pressures but also has the negative effect of lowering power.

If the ignition timing is retarded, the likelihood of detonation is reduced. The sparkplug actually always fires when the piston is still on the way up, compressing the air/fuel mixture. This is because by the time the fire starts to burn, the piston will be over top dead centre — on its way downstairs. If the comp ratio is very high, the pressure will be so intense that when the plug fires the whole lot's likely to go up in flames. Delaying the spark (that is, retarding the timing) will mean detonation is less likely.

Unfortunately, it will also mean power goes backwards! This is because the piston will have started its downward travel, so the mixture won't be so compressed when the plug set fires to it. In effect, it will be like the comp ratio's been lowered...

The best way to control detonation is by water injection.

WATER INJECTION

Way back over 55 years ago, they were using water injection on engines. These were big muthas running huge superchargers and getting great power outputs. It was all pretty urgent cos they were fighting a World War. Water injection was used on WWII aircraft during takeoff to give them that extra grunt. The supercharger boost was high — bloody high — and detonation was a real

8. Shooting a fine spray of water into the mouth of the carb will almost always stop detonation in its tracks.

9. The water evaporates and helps to cool the inlet air, much like how spraying your girlfriend's perfume atomiser all over her cools her down ... even if it has the opposite effect on you.

11. Tuning the switch-on point and the amount of water to be added can be done easily on the dyno. Here you can load the engine up, have people standing around listening for detonation, and visually confirm that the water spray system is working as it should.

10. Using a truck windscreen or headlight washer pump and an underbonnet reservoir will give you a water-injection system that works very well for well under $100. Add a switch which automatically turns on the system at high throttle openings and you're laughing!

problem. They used automatic systems to lean out the fuel and add the water, but the only problem was that if the water system broke down, the plane fell out of the air...

Water added to the combustion air has an amazing ability to stop detonation. It does this in two ways. First, it evaporates, taking heat out of the intake air in the same way an evaporative air-conditioner works. Water absorbs a helluva lot of heat in this way, so the intake air temp is considerably lowered. This reduces the temp of combustion, reducing the onset of detonation.

Second, some of the water passes into the combustion chamber in droplet form, turning to steam inside the combustion chamber. Steam acts as a anti-detonant in a way that's not really understood — but it sure does a good job. In fact, you can even reduce the amount of fuel that's being added to the air and you'll still get the same power — mebbe it's the hydrogen in the water adding to the burn. However, we don't recommend you decrease fuel — just add the water!

WATER INJECTION TECHNIQUES

Unlike with fuel, there's no requirement to keep exact ratios when adding water.

A fine spray of water at around 100-200ml/minute will control detonation to an unbelievable degree. Adding too much water will make the engine feel unresponsive and drop power. But anything up to this high level will be fine.

So how do you add the water? Easiest and cheapest is to use a windscreen washer container and its motor. Add these to a fine jet shooting the water into the mouth of the carb and Bob's your uncle! If you want to get some heavy-duty rigour into the system, source the washer and its reservoir from a truck wrecker, where man-sized bits are available for the asking. If the system is adding too much water, put a small plastic hardware-store irrigation tap in the line; this makes the flow adjustable for just 70 cents.

Switching on the water spray is as easy as using a small microswitch and a cam plate on the throttle — either at the carb or injection butterfly, or a normal switch on the accelerator pedal. Because it takes high combustion pressures to cause detonation on high-comp engines, it won't usually happen at small throttle openings. This makes it easy to trigger the water spray at throttle openings over, say, two-thirds.

It's best to have an automatic system because you won't have time to hear detonation and then activate a water injection system. For the same reason it's wise to have the water injection system operate *before* it's actually needed. Note that if you ever do hear detonation with or without the water working, back off!! Lifting your right foot can save your engine, and those subtle 'tink tink' sounds can mean the end of your engine.

Make sure you fill the water tank every time you put petrol in the car, and occasionally check that the pump and jet are working fine (nothing's blocked). If you like, you can also add to the fuel an octane booster like 104+ for extra security — but the good thing about water is it's available free at service stations.

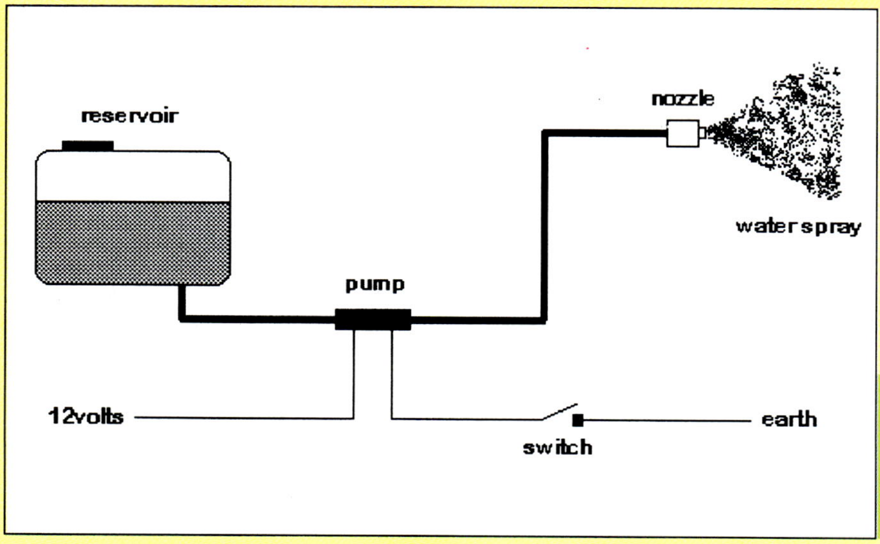

Diagram: A water injection system can be as simple as a reservoir, pump, nozzle and switch. The spray is directed into the air intake of the engine.

How To

FIT A HANDBRAKE

TO AN LH-LX TORANA

A relatively simple conversion that can be performed in one day

For some unknown reason GM-H decided to use a foot-operated parking brake in the LH/LX series Toranas. Fortunately, it decided to replace it with a real handbrake lever between the seats on the UC model. You can convert the older cars to the UC system.

Our friendly local wreckers, Sims Spares of St Peters, had a load of UCs in stock. They cut out the tunnel mounting area and supplied the entire console, brake lever and rear brake setup for a hundred bucks.

The entire conversion can be performed in one day with little more than hand tools. You will need an oxy set or MIG welder to replace the centre section of the transmission tunnel (and someone who knows how to use it).

The section of transmission tunnel that is required has certain brackets and attachments that need to be left intact for correct installation and operation. The only differences in the floorpans would appear to be three bolts sticking up from the floor to bolt the handbrake onto, and two oval-shaped holes for the cables to route through. If you think you can just weld the bolts onto your floor and cut the cable holes, think again.

Any structural modification should be performed only by competent people. Consultation with an approved engineering signatory is recommended. I feel that the methods described and demonstrated in this feature would satisfy an engineer's requirements.

Before cutting the panel from the donor car, careful measurements should be taken and the section of tunnel should be cut out with plenty of space left around it for trimming later. The accompanying pictures show you how big the tunnel section was when cut from the donor car, in this case a four-cylinder UC Sunbird.

The underneath shots show the various brackets, fittings and braces needed for strength and correct operation. You will also note that the cutout section included the seatbelt mounting plates. Do not remove these mounts from your floorpan and replace them with those from the donor car. It would have been slightly easier to leave them in situ and replace the tunnel section at floorpan level but it is not wise or legal to alter or change seatbelt mountings without prior engineering approval.

In addition to the structural bits you may also need the support brackets for the UC consoles if you are intending to use them. We strongly recommend that the rear half (handbrake cover) part of the console be used at the very least as there are sharp, protruding sections of the handbrake mechanism that can injure you in the case of a prang.

We have replaced the entire rear brake assemblies with those of the UC. This was done because the handbrake cables are different from the original. This conversion includes the backing plates and wheel cylinders that have a metric thread on their hydraulic lines. We also obtained the brake hoses and fittings which attach to the diff (included in the cost) to ensure correct connections.

Story: JAMES GARROD

Pix: JAMES GARROD & MAUREEN FIDDLER

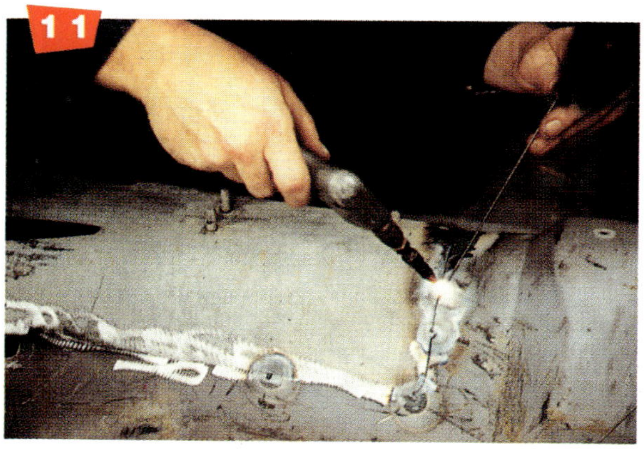

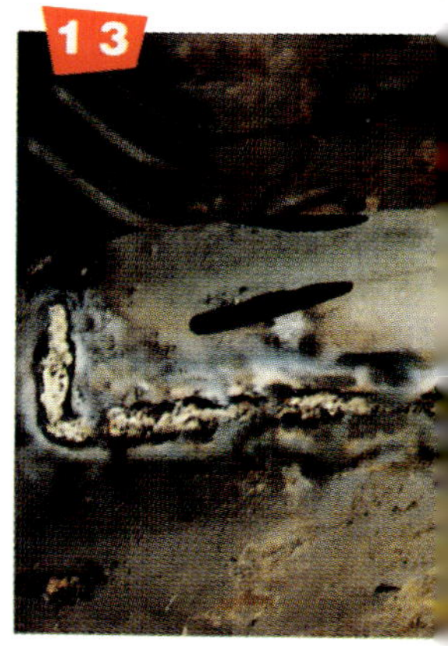

1. This is the size of the hole in the donor car when the bits are cut out for this conversion.
2. What is needed for the conversion: complete rear brake assemblies including cables, tunnel section, lever, brackets and linkages. Note the small brackets in the centre of the cables — they're to bolt the cables to the body to prevent flex and damage.
3. Close-up of removed tunnel section, showing lever, mounting bracket, cables and floor grommets. Yellow wire is for handbrake-on warning light.
4. Close-up of trimmed, ready-to-fit tunnel section. Note the reinforcing under the tunnel for the handbrake lever and the cable locating brackets behind the grommet holes. These are an integral part of the handbrake's operation and must be fitted.
5. Thoroughly clean the floor area and remove all remaining underfelt, sealants and oil from above and below the welding area.
6. Trial fitting the tunnel section with lever in place. Marked initially in green, but on closer inspection it was found to be 15mm too far forward. Dimples cut in edge of new panel are where the two upper spot welds were on original seatbelt mount. These were lined up with existing belt mounting spot-welds to ensure correct location.
7. Green outline was the first — but incorrect — positional. The first black line is the correct positioning outline and the second line is an undersize cutting line. This sort of panel replacement should never be an exact-fitting, butt-welded job. The panel overlap provides additional strength to this area after welding due to the increase in metal thickness. Careful cutting with a jigsaw is the best method for a smooth, distortion-free cut.
8. Completed cut showing the extra metal left around the seatbelt mountings. If required by an engineer, this extra metal can be welded to the tunnel from underneath for additional strength

in this area. We left the extra metal to absorb heat from the welding/brazing, thereby reducing the chances of molecular structure changes of the mounting metal.

9. Metal area surrounding the weld is cleaned of all paint with a grinder. New, clean, bare metal makes for a better weld.

10. Now the headaches begin. The UC Torana has a slightly different floorpan shape in front of the rear seat. L34s and A9Xs have the same floorpan shape which is designed to provide clearance for the Salisbury diffs in these models. This picture shows the new panel tack-welded into place front and back. Careful massaging of the rear floorpan area combined with some slight heating and bending of the rear section of the new panel allow it to be fitted correctly.

11 & 12. Tack-welding the new panel into place locates it and allows relatively distortion-free welding/brazing. A MIG welder is best for this purpose due to its low-heat method of welding.

13. This is where the most stress will be placed during operation. We tack welded the rest of the panel. The rest of the panel was brazed in place around its complete perimeter. The brazing is strong enough for this relatively low-stressed area and has the added advantage that it will never rust along the joints. A further advantage is the lower distortion due to the lower heat used for brazing.

14. Fitting the UC rear drum assemblies to the LH axle. They will be totally rebuilt despite being good enough to be connected and used as they were.

15. Fitting of the axles takes some care so that splines and housings are not damaged. Note the newly fitted bearings and seals — you might as well replace them while it's all pulled apart.

16 & 17. The axle retaining nuts are tightened and the drum is fitted; that finishes the wheel end of things.

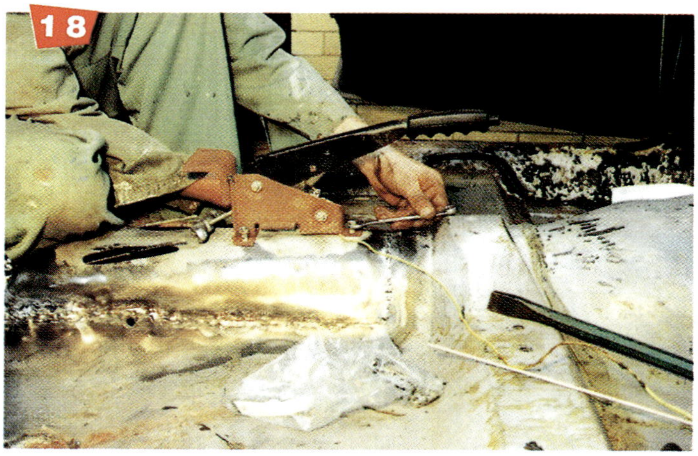

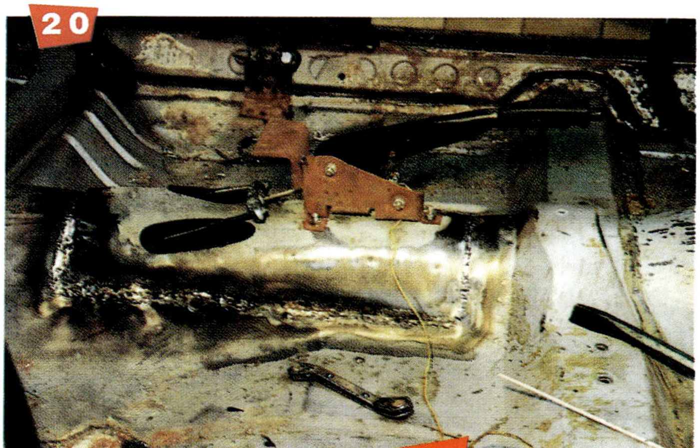

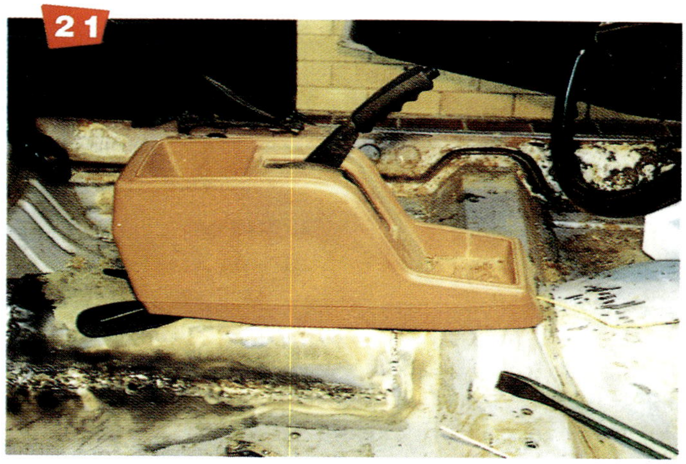

How To

18. The handbrake lever can now be bolted in and connected to the cables, but only after the outer cables have been correctly located in and through their brackets.

19. Correct fitment of the cable grommets is necessary to stop fumes, oil and water from entering the cabin. Once they're correctly positioned and seated, a thin layer of contact glue or Silastic should be squirted around them.

20. An excellent range of cable adjustment is provided by this setup. Even after individual brakes have been adjusted at the drums there are still the lever cable adjuster and individual cable adjustments as well.

21. Console being trial fitted. Of course, the carpet and underfelt is yet to go in.

Coming to GRIPS

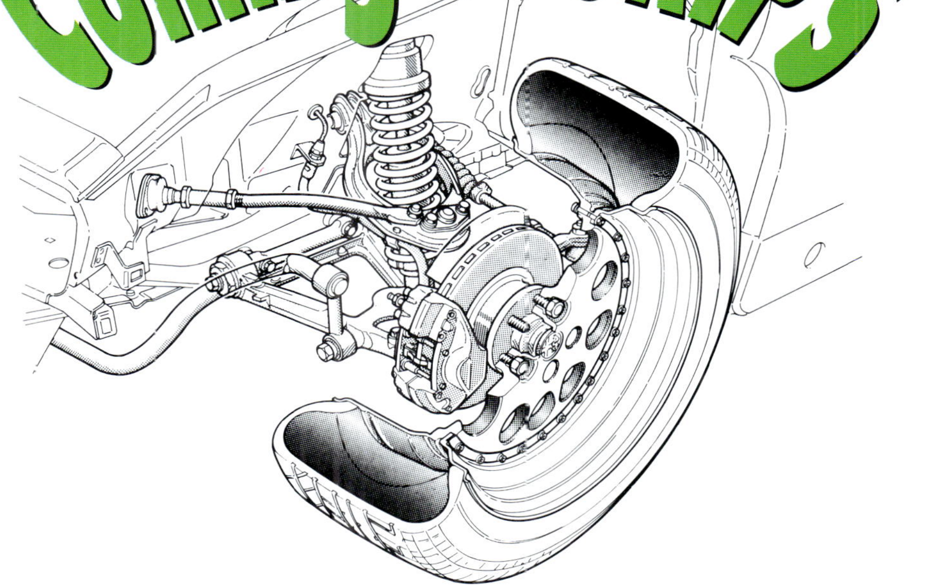

The job of the suspension system is to get a solid hold on the road. We examine the various suspension components

Australian suspension engineers are very highly regarded worldwide. There are several reasons for this. The typical Aussie driver likes a responsive car which can be hustled hard at corners, no matter how rough, with no fear of it spearing off into the bush.

And Australian roads vary tremendously in quality. We have freeways which are among the best in the world and we have dirt tracks that are among the worst. Between these two extremes there are moderately good country roads, badly broken bitumen bush tracks, rough and ready suburban streets — and a dozen more examples of the ancient art of road making and breaking.

To design a suspension system to handle all these conditions is impossible; by necessity, the word compromise frequently creeps into conversations with suspension designers and component manufacturers. Compromise not only in the need to handle diametrically opposed road conditions but also compromise in having to keep components relatively simple, tough, easy to fit and moderately priced — all at the same time...

As well as suspending the car, the suspension's task is to assist with steering precision, roadholding and comfort — all at the same time.

To give you an idea of what suspension design is all about we will look briefly at each component and describe what it does. In future features we will go into each part of the all-important suspension system in considerable detail.

Springs

If roads were perfect cars wouldn't need springs, indeed they wouldn't need much at all in the way of suspension. Springs exist to absorb road irregularities — natural irregularities like potholes, dips and bumps, and unnatural irregularities like those bloody awful speed humps (shouldn't they call them slow humps?) which seem to be creeping in everywhere.

Springs can take the form of leaf springs, which are good for carrying loads but don't react quickly enough to give good handling characteristics. Leaf springs haven't been seen on the front of many cars since the 1940s and went out on the rear suspension in the 1970s and early '80s. Some four-wheel drives and many trucks still use them but their drivers aren't particularly interested in capable cornering characteristics.

Coil springs are almost universal in modern cars, in both the front and rear suspension. Holden went to rear coils with the HQ, Ford with the XE Falcon. Coil springs give a softer ride than leaf springs. They have reasonably regular characteristics over a greater length of spring travel than do leaf springs, and they react much more positively and quickly, thus permitting faster cornering speeds with more control and safety.

Torsion bar springs are virtually unknown in modern cars; Chrysler stuck with them for longer than most other car makers.

Naturally, there's an infinite number of spring stiffness settings on the market. Stiffness not only in compression — when the car hits a bump and the spring is compressed — but also in rebound, which is what the spring does after it has been compressed.

Shock absorbers

The first thing we have to say about shock absorbers is that they DON'T absorb shocks. The name is wrong — their correct title is dampers, or spring dampers. Dampers exist, as the name suggests, to damp out the movements of the springs.

If a car had springs but no shock absorbers (let's bite the bullet and use that name) it would bounce up and down almost uncontrollably every time it hit a bump or hole and the springs alternately compressed and rebounded. And would keep on bouncing until the springs eventually settled themselves down — by which time the car would probably have hit another bump and the whole commotion would have begun all over again…

We won't get too deeply involved in shock-absorber technicalities at this stage (we'll do a lot more in future issues), except to say that shock absorbers are partly filled with oil and use the friction and speed of oil movement through a series of orifices to damp out oscillations within the shocks, and therefore the springs.

In some shockies gas pressurisation is used to help the oil do its work without generating excessive heat. Gas-pressure shocks are particularly useful if the car is used in circuit racing or in rough road conditions where there is constant movement of the shocks — there's little advantage in using them for street cruising.

The rate at which a shock absorber reacts both in compression and in rebound influences its effect on comfort, road grip and steering precision. More expensive shockies can be fairly readily dismantled to have different valves installed in them to alter their characteristics. Some shock absorbers even have external adjustment.

Swaybars

Again the name is wrong. They should really be called anti-swaybars. To further confuse things, swaybars are also known as rollbars and anti-rollbars.

Basically the task of a swaybar is to give some communication between the suspension on one side of the car and that on the other side; either front to front or rear to rear.

If the car is running in a straight line the swaybar doesn't do much, but when the car corners the bar resists body roll and gives the tyres, particularly those on the outside of the bend, which are doing most of the hard work, a closer-to-level grip on the road. The idea is to make the car turn in cleanly and come out of the corner with a minimum of trauma.

Typical Steering and Suspension Systems

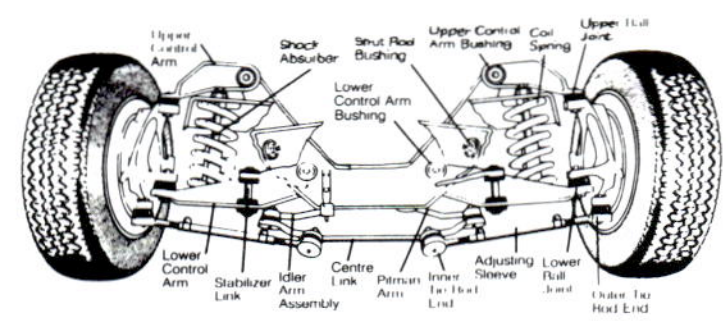

Conventional Steering and Suspension with Coil Springs

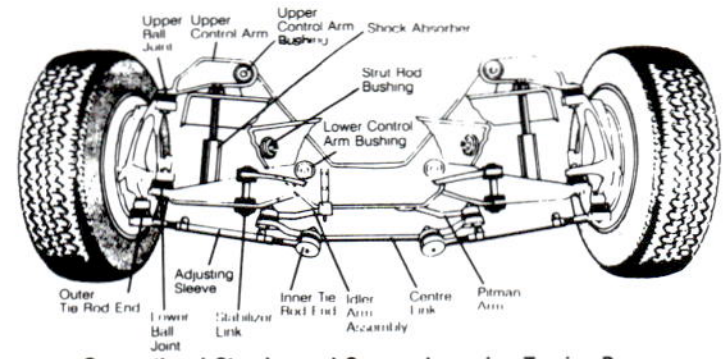

Conventional Steering and Suspension using Torsion Bar

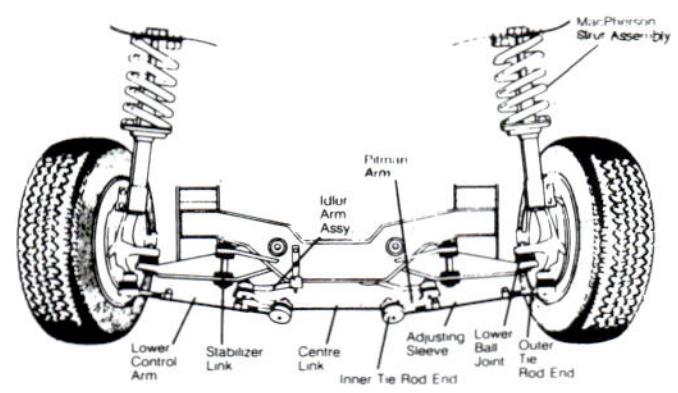

Conventional Steering with MacPherson Strut Suspension

If the car hits a bump with equal force on both sides the swaybar doesn't do anything, but if the car hits a bump (or hole) at one side only, the swaybar resists the rise or drop of the suspension on that side and helps to hold the car closer to level.

As with everything else in suspension design, there are varying degrees of rigidity in swaybars. Bigger ain't necessarily better. Too much stiffness in a bar not only stuffs up the car's comfort level but may also ruin the handling. A car needs some body roll in order to give the tyres on the outside of the corner some bite onto the road surface — too stiff a bar can cause the car to skate sideways and create real problems.

Control arms

Also called suspension links, these are the components which connect the springs, shock absorbers and swaybars with the car's body. The design and mounting of the arms affects the way the wheel angles change in cornering and when hitting road irregularities. To a lesser extent they also control suspension movement under acceleration and braking.

You get the control arms and their mounting points when you buy the car. Changing them is a major job and one not normally undertaken in a street machine rebuild. Drag racers and, to a lesser extent, circuit racers may wish to alter the control arm type or setup to gain better traction.

The job of the control arms is a complex one because they are almost invariably moving in two directions at any one time. If you hit a bump it moves up and back; hit a hole and it moves down and forward.

If you are cornering and accelerating at the same time as you hit the bump then the suspension is doing four things at once: it's resisting sideways movement, being forced back by acceleration, forced up by the bump and probably back by the bump. All this is happening at the same time — and keeps on happening at various stages to differing degrees as the car continues along the road. That suspension under your car is probably doing a helluva lot more than you realise…

Oh, and the control arms on one side of the car are also getting input via the swaybar from the other side of the car so they are moving to a lesser extent according to what's happening on two sides of the car at the same time.

Complicated, isn't it? Even in today's

Typical Steering and Suspension Systems

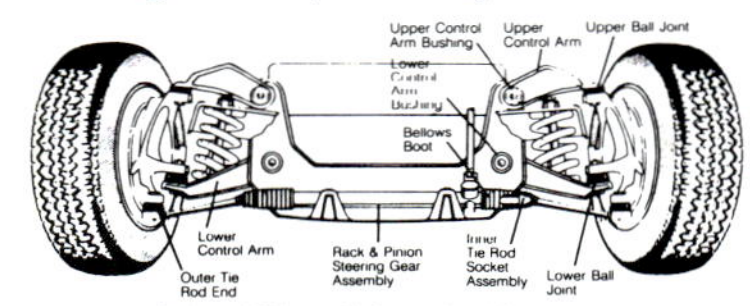

Rack and Pinion with Conventional Suspension

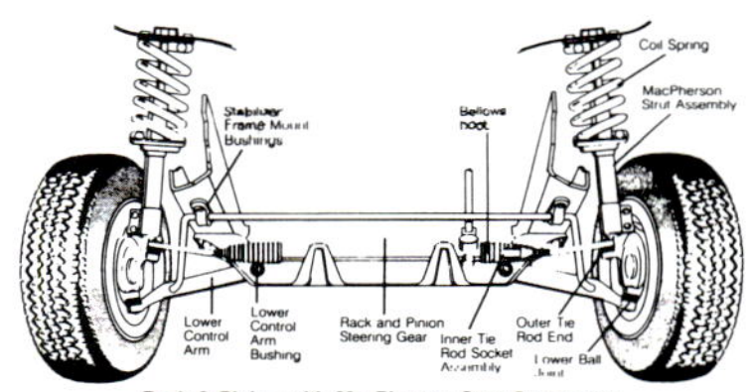

Rack & Pinion with MacPherson Strut Suspension

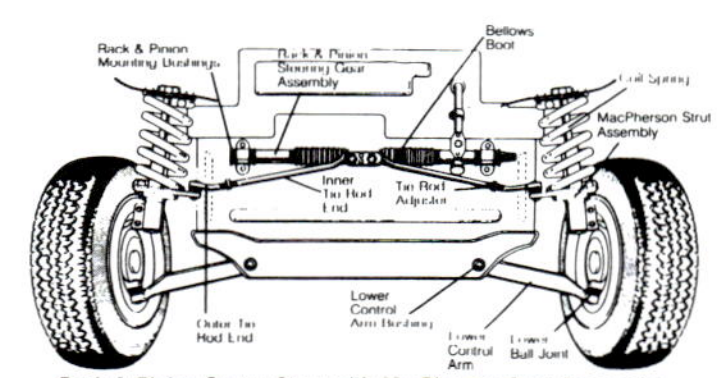

Rack & Pinion Centre Steer with MacPherson Strut Suspension

Story: EWAN KENNEDY

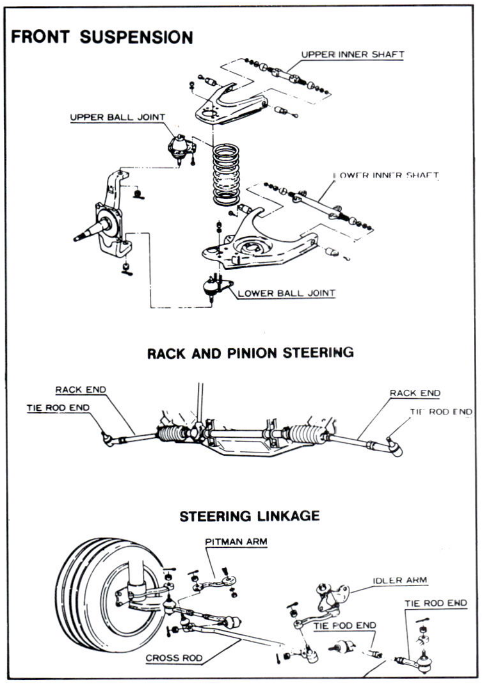

highly computerised environment there is still an element of art in the science of suspension design.

Front-end alignment

The front-end alignment, or steering geometry as it's sometimes known, plays an important part in the way a car handles. The main angles which affect a car's handling are camber and caster.

Camber is the angle at which the bottom of the tyre sits further out than the top, or vice versa. Take a look at the front wheels of today's touring cars to see some really exaggerated negative camber angles. These put the outside front tyre squarely on the road for maximum grip when the car is leaned over in a corner.

Naturally, running a car at big camber angles causes uneven tyre wear. That's OK on a race car but may not be acceptable in real-life driving.

Caster is the angle at which the wheel trails behind the centre of the axle — look at the casters on a shopping trolley and you get a good example of how the wheels act. Pick a good trolley (if you can find one) as damaged trolley casters carry on with some peculiar antics. Caster's main job is to keep the car running stably in a straight line; it has a marked effect on steering feel and response in a corner.

If you are on a tight budget then simply modifying the caster and camber angles (and toe-in or toe-out — more about these in a future feature), or even just setting them back to the car maker's specs, could make a noticeable improvement to the way your car handles.

Ball joints and bushes

Everywhere that movement takes place between two components in a suspension system there are ball joints or bushes to prevent metal-to-metal contact.

Ball joints allow movement in several directions, bushes in one direct on only (actually, bushes permit very limited movement in a second direction as well — more about that in a future issue).

The material from which the suspension bushes are constructed has a much greater effect on the operation of the complete suspension system than you might realise.

Simply replacing the ball joints with new ones and changing the suspension bushes may go a long way towards giving you the sort of feel you like in a car's handling and ride.

Naturally, most ball-joint and suspension-bush changes take place at the time the springs, shocks and bars are done, but you might like to experiment with a bush change on its own if your budget is on the slim side.

More to follow

The aim of this feature has been to give you a brief overview of what a suspension system does. Hopefully, you have come away with a solid appreciation of the complexity of design, and a feeling for how hard the suspension is working, even under everyday conditions.

There are many myths about how a car's suspension can be improved; mainly these revolve around having the lowest ride height with the stiffest springs, bars, shocks and bushes. If you are simply out to impress your mates and are willing to tolerate an atrocious ride, then go for it.

But if you want to have a car with tenacious road grip and which is safe and a pleasure to drive and to ride in, then keep reading our continuing series of suspension articles for all the good oil ... and the good metal, and the good plastic.

Thanks to the following companies for their assistance in compiling this feature:

East Coast Suspension (02) 545 1044
Fulcrum Suspension (07) 892 2647
Jim Hunter's Suspension (02) 647 1544
K-Mac (02) 526 2744
Monroe Shock Absorbers (consult Yellow Pages for nearest outlet).

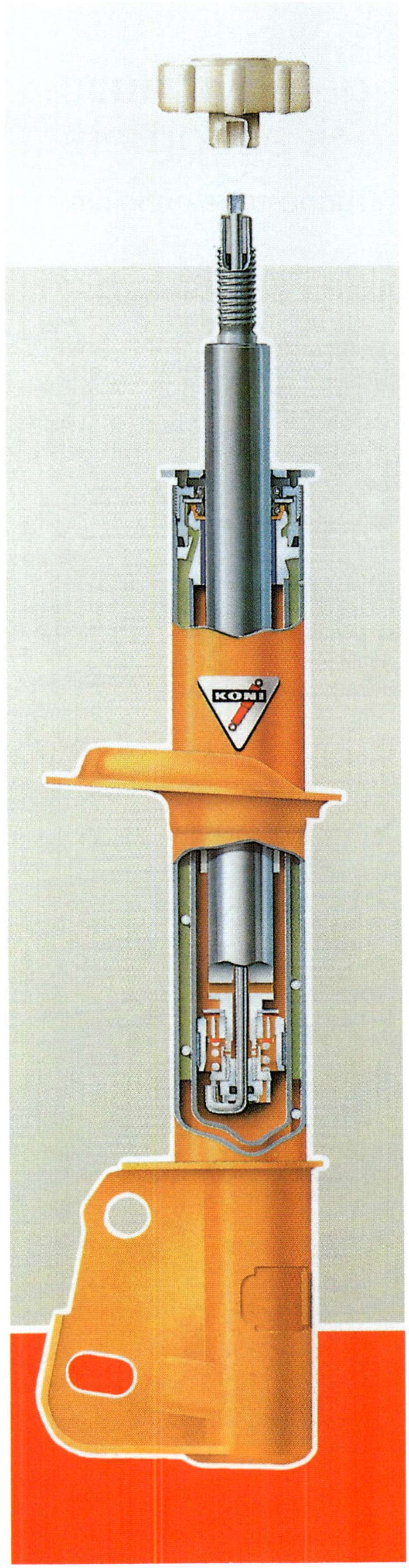

ENGINE BAY DETAILING- alternators

The days of chroming everything that'll unbolt are all but gone — but these days there are plenty of other options

Fads and fashions come and go. Currently we're seeing more careful detailing and greater presentation under the bonnets of streeters and show cars — once the last place to get attention except for the go-fast goodies.

Now the wiring is hidden, batteries are relocated, radiators shrouded and hoses braided. Each and every component gets the treatment.

The options are many when it comes to detailing your engine bay, so you should plan your attack, co-ordinating colours, making use of contrast and generally creating an attractive and efficient package.

Chromeplating has seen a decline these past few years but other electroplating processes — such as anodising with its range of colours to match hose-clamp fittings — have come to enjoy extensive application. There are even paints available to match the colours used in the anodising process.

Polished alloy is still with us, of course, but now the smoothed surfaces are often covered in layers of colour as part of an overall scheme. A good example is Ron Barclay's multi-award-winning HQ ute which has had many surfaces smoothed to a polished finish then detailed with paint to match the body colour.

Then there's powdercoating, which offers a very durable surface at reasonable cost, though it does have a few limitations. It's certainly not something you can do at home; the range of colours is still light years away from what's available in auto paints; and you cannot prepare a perfect surface as easily as with durable paints.

So let's get dirty and get detailing.

The alternator is a vital part of the electrical system but it sure can be unsightly perched at the top of your engine bay. At least those readers who run Chevy engines can often use aftermarket bracketry to mount the alternator lower, closer to the chassis rails — almost out of sight, in fact.

Here we show you how to tear down a typical alternator, then prepare and paint it in a manner that's applicable to most components under your car's bonnet — or anywhere else, for that matter.

If you choose to use paint as your medium, remember there are many parts that are sensitive to dirt, grit and paint. Such areas must be masked to prevent contamination.

Follow these steps to disassemble your alternator yourself, but by far the easiest method is to take it to an autoelectrician for an overhaul, ask him to give you the parts to detail, then take back those parts so he can complete the reconditioning process. Hopefully, it won't have to come apart again!

Remember: when you use paints of any kind, always follow the manufacturer's instructions *to the letter*, especially the safety measures.

Now it's time to save some bucks and achieve show-worthy results — not to mention the satisfaction you'll get from doing it yourself!

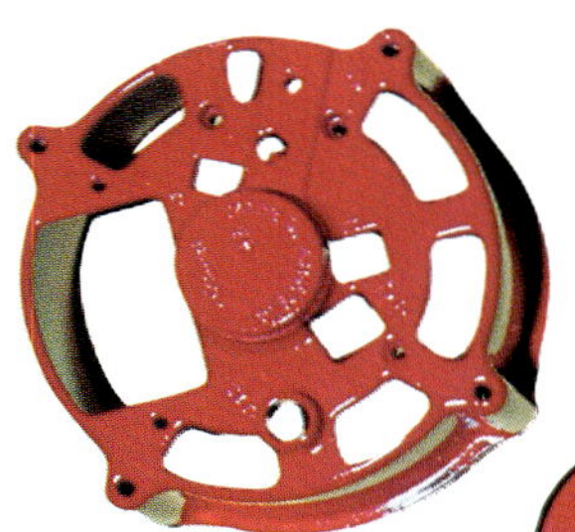

SOURCE:
Slee's Electrical Service
Ph: (059) 75 2900

story/photos: **Rob Warden**

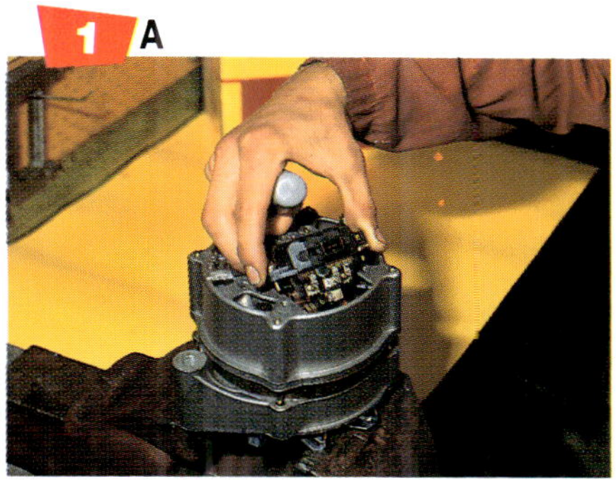

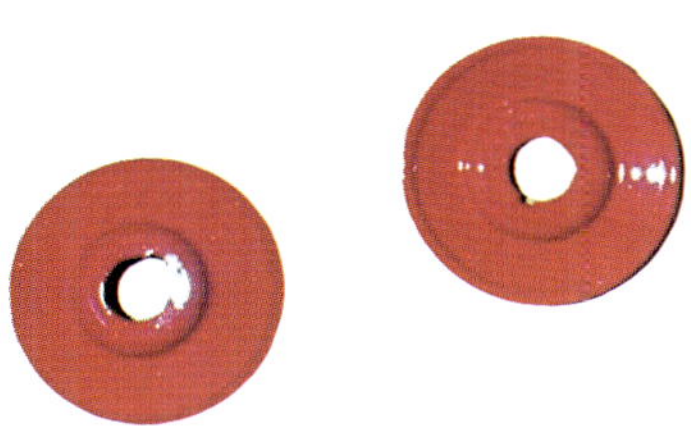

1&1A. Start the disassembly process by carefully clamping the alternator in a vice. Remove the two screws to take out the regulator.
2. There are four long screws that join the alternator ends. The ends may be located in four different positions as the casing can be rotated. Scribe your alternator so that you can reassemble it accurately. (Be careful not to lose the scribe marks during the detailing process, however.)
3. When the four long screws are removed, separate the alternator carefully.
4&5. emove the nut that retains the pulley and fan. Make notes on the sequence of the parts removal.
6. Open the jaws of the vice and use it as a stopper, pushing the rotor shaft out of the casing with a soft hammer.
7. This is how it should look at the halfway point.
8. On the other casing end, remove the four screws that retain the bearing.
9. Remove the battery power(+) terminal.
10. Unbolt the suppressor and remove the terminal.
11&12. Most alternators have three screws that release stator windings and rectifier plate from the casing.

13

14

15

16
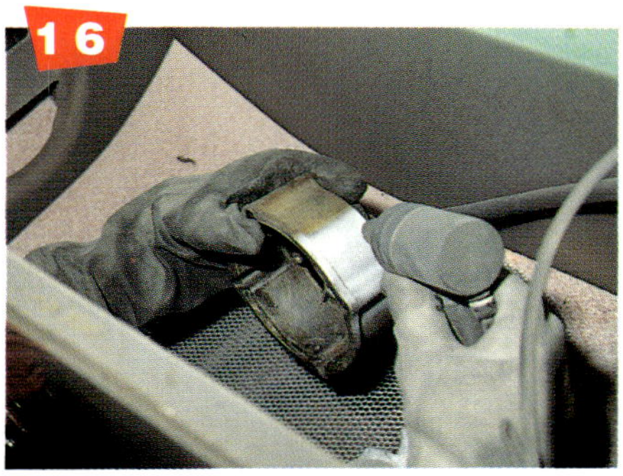

17

18

19

20

21

22

23

24

13. These are the components we will be detailing: the two casing ends, the pulley and washer.
14. If you are detailing an old alternator it may look more like the grungy sample at left than the shotblasted item at right.
15. Start by cleaning thoroughly with solvents.
16. Shotblasting or sandblasting with a fine grit is ideal preparation in the clean-up stage.
17. You can see the area at left is very dirty and stained while the blasted side at right is OK for painting, providing an etched surface which is excellent for paint adhesion.
18. Rather than paint, you may prefer to polish the alloy parts.
19. Even after brief polishing the alternator casing is starting to sparkle. The blasted surface is best suited for a paint process.
20. Shown here are flat, curved and 'rat-tail' files that can be used to smooth and debur the alloy parts. Practise using the grinding stones before you try them on the real thing.
21. Carefully and gradually file the highs away; any low areas can be filled with a two-part filler.
22. After filing, filling and sanding, you're ready to apply the primer. Thoroughly clean the surfaces with thinner or grease remover. There must not be any impurities on the surfaces to be painted.
23. As in all painting, technique governs the final results. Follow the manufacturer's instructions on mixing and application. Be sure to wear the appropriate mask for the type of paint to be used. We suspended the parts from a plank supported by two ladders; each part hangs by fine welding wire.
24. Once the primer is dry, carefully sand smooth and use a 'tack' rag to remove sanding dust, etc.

25

26

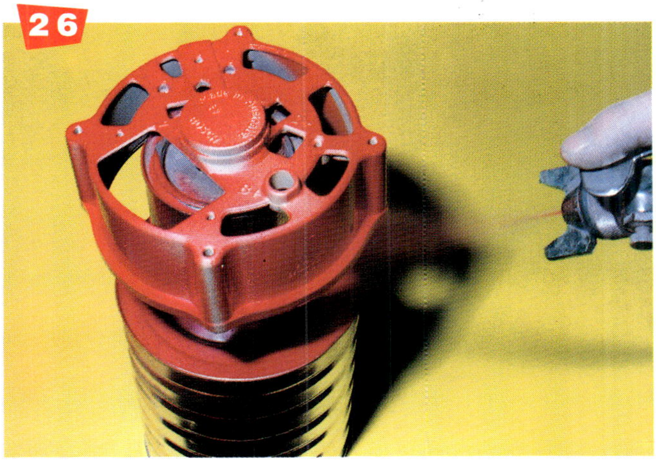

27

28

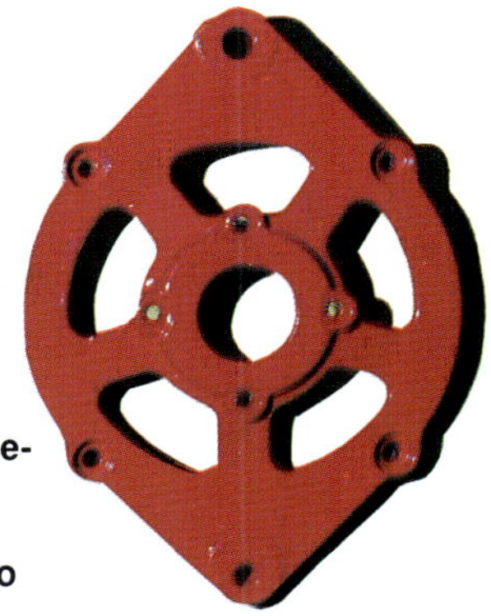

25. Only a small amount of paint is required to coat these alternator components, so be careful not to mix too much. We used quality Dulux 2K products for an excellent off-the-gun finish that's durable and resists solvents. You can also use an over-the-counter touch-up spraycan of acrylic lacquer then polish it to a gloss.

26. Instead of suspending the parts to be painted you can mount them at an accessible height. Many paints need a light 'tack' coat to support the next coat, building up the coverage. Leave the parts to dry thoroughly before contemplating reassembly.

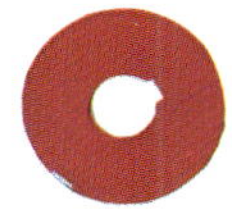

27. The dry painted parts are ready for reassembly. A similar finish can be achieved by powdercoating but the method we have used costs only the price of a couple of spraycans.

28. Refer to the disassembly stages shown to put the alternator back together. This is an ideal time to have it reconditioned by an autoelectrician; that way, he handles the teardown and assembly stages, then does a test on the alternator before you refit it to your engine.

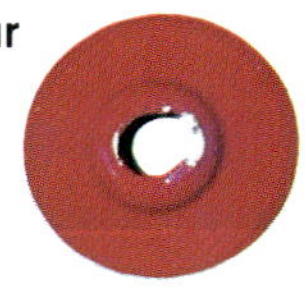

THink Tank

How To build a drop tank

Custom fuel tanks come in all shapes and sizes and are especially important for street rods. Owners of pro streeters almost invariably follow this route, too, mainly because the original tank almost certainly won't fit with revised rear chassis.

Most things in car construction are simple enough if you know what you're doing, and building a new tank is no exception. These are the things you'll have to think about: shape and material to be used, internal baffles, inlet and outlet pipes, the fuel gauge sender unit, how you're going to mount the tank and protective treatment to stop it corroding.

SHAPE

Although a simple box shape is the easiest to make, will it fit in the available space? Irregular tanks aren't necessarily difficult to make and, in most cases, size constraints will force the issue. With odd shapes, remember that if the base isn't flat the fuel pickup pipe has to run to the lowest point.

You can work out capacity using the following formulae: 1 litre = 1000 cubic centimetres (so 5 litres = 5000 cubic centimetres); and 1 gallon = 277.3 cubic inches.

MATERIAL

Your selection will be governed by what you want and what you can work with. Here's what there is to choose from:

Mild steel is heavier than aluminium but stronger, gauge for gauge, so it's harder to form and bend. It can be oxy or arc welded but distorts more. It will need protective treatment to prevent rusting and fuel contamination.

Stainless steel is similar to mild but offers superior appearance and doesn't need anti-corrosive treatment. It can't be oxy welded (except by brazing); ideally it should be arc welded. Its relative hardness also makes it difficult to work with.

Aluminium is usually the favourite because it's easy to work and light. It can be oxy or arc welded.

Carbon fibre is used in the race car industry, but always with an internal bag or bladder.

Plastic appears extensively in new cars because of its low cost, light weight and ease of mass production. It also resists fractures well in the event of impact. However, it's unlikely many production car tanks would be suitable for our use and they can't be modified.

CONSTRUCTION

Melbourne-based Price Engineering fabricates a huge range of tanks and ancillary components. They've shown us how they produce an aluminium HQ tank.

The design of this tank is simple enough, with the main sections produced from a single sheet of aluminium.

The first thing to note is that when folding sheet metal to produce a particular-sized tank, you need to allow for the material taken up at the bend. This is referred to as 'bend allowance' and can be calculated as twice the material thickness per right angle bend. Material thickness (SWG) will usually be 16 or 18 gauge when using mild steel or aluminium, and 20 or 22 gauge for stainless.

Price Engineering always uses a folding press, but if you don't have access to one, you can enlist the help of the bench in your workshop. Simply clamp the metal to the bench using a sturdy piece of box section or angle iron over the top. By aligning the metal with the desired bend line, directly over the edge of the bench, you can carefully bend it

Story/photos: TREND SETTING PICS

down to the required angle. This method won't give you a perfectly clean bend, but if your tank is no longer than 60cm it'll work acceptably.

Before committing yourself to bending the main section of the tank it's a good idea to bend up a 25mm-wide test piece to check your bend allowances. This is also very useful for test fitting in the car — better a mistake now than later. At this stage the main section can be tack welded together.

INTERNAL BAFFLES

With any tank over 20 litres in capacity it's advisable to stop the contents from sloshing around too much when driving. This can be achieved in two ways.

One is to install baffles or 'walls' which simply split the tank into smaller compartments (but you face the possibility of internal welds breaking and creating stress points).

The other is to fill the tank with a rigid sponge-like material. It has the added advantage of helping prevent explosion in the event of accidents and/or fire, but it's quite expensive and complicates sender unit fitment.

INLET AND OUTLET PIPES

It's simple, isn't it? One hole for in and one for out? Usually. But there are exceptions.

First, as fuel leaves the tank, air must take its place either through a vented filler cap or a separate breather pipe which lets air in, but no petrol out. This is achieved by fitting a non-return valve in the pipe or by looping the pipe high up and back down to floor level. The disadvantage of the latter is it puts a potentially dangerous fuel pipe inside the vehicle.

The main inlet pipe may carry the fuel filler cap directly or connect to it via a short hose, depending on the type of cap fitted.

The outlet pipe can be fitted directly into the base of the tank or via a siphon pipe fitted in the top but leading down to the bottom (slightly above to prevent picking up any debris).

A base fitting gives a self-priming supply of fuel any time the tank is filled after being emptied. The drawback is that the supply pressure varies depending on how much fuel is in the tank, and some

1. Removing aluminium sheet from stack. Price Engineering always uses 5251 marine grade aluminium (2mm) as it doesn't stain. The more normal 5005 grade may stain but is cheaper.

2. Marking the dimensions prior to bending.

3. Cutting the sheet to size on the guillotine.

4. Using the correct folder blade (a radiused edge as opposed to standard blade) drastically reduces the orange peel effect.

2

1

4

3

5

6

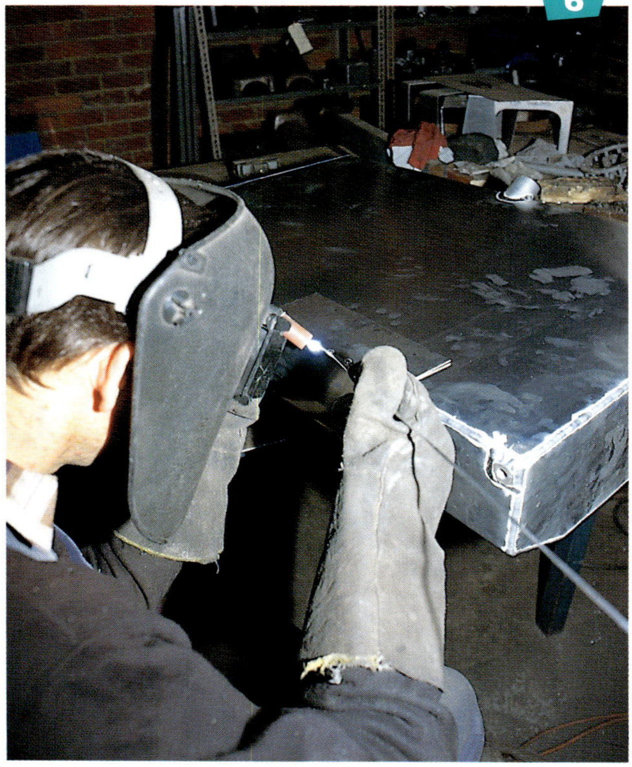

7

How To

5. **First fold. The Price-modified folder provides for one-piece construction, therefore requiring only one weld.**

6. **Welding the tank using a TIC welder.**

7. **All Price Engineering Holden tanks use factory-type cast fittings which accept factory fuel pickup/sender units. The casting is welded in place.**

pumps and carburetors don't like this. Another problem is that the tank has to be completely emptied prior to removal. A siphon pickup doesn't suffer from pressure change; neither does it require the tank be drained for removal.

You may have to fit one last pipe: a return for excess fuel. Fuel-injected engines always have this; most carburetor engines don't need it.

The main outlet pipe (and line to the fuel pump) should have a minimum of bends, and any which are present should be smooth, with no kinks; otherwise cavitation will result and cause a drop in pressure to the pump. For high-demand engines, the resulting lean mixture can be disastrous.

You should also consider the size of fittings used in relation to the thirst of an engine. In many cases a thirsty Chev or Cleveland will need half-inch fittings.

From a safety point of view all pipe ends should be flared to prevent the hose from pulling off, unless you're using Aeroquip screw fittings.

FUEL GAUGE SENDER UNIT

There are two basic types — the old-fashioned swing lever and the newer dip-pipe design. Your choice will often be governed by a desire to match the fuel gauge to your other instruments.

The range length of this unit obviously needs to match the tank depth while fitting the flange with either a screw fixing or a ring of bolts. Either way, a suitable flange needs to be machined. In the case of our HQ tank, a standard factory sender and pickup is used.

MOUNTING

The first thing to remember here is that fuel weighs exactly one kilogram per litre, about 10 pounds per gallon. A look at production vehicle tank mounting techniques reveals two main methods.

Flange type: Pressed tanks tend to be made from an upper and lower pressing, both with a flanged return. These returns are then seam welded. It's an easy way to join the two halves and can be duplicated in custom-built tanks, but the resulting flange may not look the goods.

With **strap fixing**, the tank itself is not fixed but is secured in a suitably shaped recess by straps (usually two).

Traditionally, most drop tanks are supported by straps, whereas race tanks must be secured via a much sturdier arrangement.

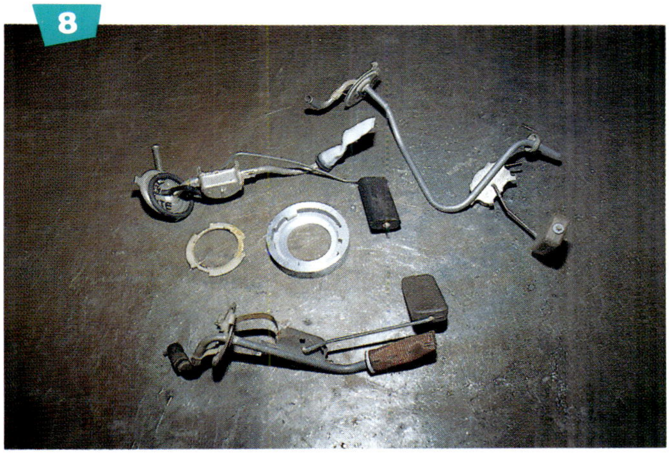

PROTECTIVE TREATMENT

The kind of protection you need will depend on the material. Mild steel, for example, will go rusty inside. Electroplating is a good way to prevent this and both zinc plate or galvanising work well; you can use chrome if it can be seen and you want to show it off.

Aluminium will corrode but anodising remedies this and can be coloured if you so desire.

You may decided to paint as well as, or instead of, plating, but the latter is the only sure way of protecting the inside.

Finally, make sure you use real fuel hose — either Aeroquip or rubber hose intended for petrol. Usually this has 'Fuel' printed along it. Beware of cheap fake stuff; it just isn't safe.

Thanks to Price Engineering. Their phone number is (03) 354 5307.

8. A range of standard Holden pickup/sender units surrounding custom Price fittings.

9. An HQ Holden race tank comes complete with custom filler flange incorporating filler, breather and pickup.

10. Race tank is fitted with bladder and foam baffle.

11. Petrol-resistant foam almost completely eliminates surge — it's the best baffle and only takes up 2-3 per cent of tank space.

12. Just a small sample of Price Engineering tanks.

13. The tank being fitted under an HQ Monaro coupe.

2

VR Commodore
Our choice of Weapon.

AIRBOX A

After watching any Formula series racing you can draw the conclusion fairly quickly that there is obviously something valid in inlet air box design. The current generation of Formula One airboxes sit proudly up above the driver's head directly in the airstream, ready to effectively 'grab' whatever air comes their way.

The concept of these airboxes is to increase engine horsepower by providing a cooler and denser inlet charge to the engine, and also by forming a higher than positive pressure in the inlet manifold at high speeds. That's right, at extremely high speeds a normally aspirated car can actually develop boost just by the ram effect present with a correctly designed forward facing airbox.

There is a simple equation that governs this law to inlet ram effect that has been used by racers in the know for years, allowing them to calculate possible gains with this type of airbox. It is:

$$\text{Intake ram effect (psi)} = \frac{\text{air density (lb/ft)} \times (\text{air velocity} \wedge 2)}{288 \times g}$$

This equation assumes that the air inlet duct is 100% efficient, which will obviously never happen but provides us with an accurate guideline. Substituting some numbers gives:

At 100mph intake ram = 0.076 x (150 ^2) / 288 x 32.2
Intake ram = 0.18psi

At 140mph intake ram = 0.076 x (207 ^2) / 288 x 32.2
Intake ram = 0.35psi

Up in the air and ready for our Airbox mods.

3-inch flexible tubing is used to duct the air.

By Rick O'Toole

DVANTAGE

In the quest for extra power, there are always the same key elements that people try to improve, whether it be inlet-related, exhaust-related or even inside the engine itself. There are very few external modifications that can be performed on any engine in any car that fall into the description 'free horsepower' — in other words, power gain without cost in driveability or economy. Airbox, or ram airbox, design is one of the only free areas of power increase I know of and the strange thing is, not too many people take advantage of the possible gains to be made. Let me explain...

AIRBOX ADVANTAGE

4

More tubing has a wire insert to keep the sidewalls stiff.

5

6

It's generally a good idea to have some sort of stopper. These spun ally units gauze mesh are perfect.

To calculate our pressure gain over standard it's best to remember the rule of absolute air pressure stating that at a static condition at sea level air pressure is 14.7psi. Now if we can increase that air pressure by 0.35psi to 15.05psi our total air pressure gain is 2.3%.

Another important point to consider is that for every degree cooler we make the air temperature of the inlet — and being a ram effect airbox with the pickup facing the incoming airstream — we stand to gain roughly an extra half horsepower of engine output. In the instance where we cool the air charge by another 10 degrees and are moving along at 140mph we could gain an extra 5-8kW of power just by designing our air pickup properly.

As you can see, the real gains to be had with a ram air system are only apparent at speeds greater than the legal limit — at 60km/h the effect is minimal and would only represent in the order of 1-2kW, even taking into account the possibly cooler inlet charge. However, also bear in mind that any free power is good power and must be had!

An important point I have not covered earlier is the location of a ram effect pickup. This types of air catcher must run directly in line and facing towards the source of airflow (ie the front) of the car at all times. From the pickup point — that should be flush with the body mainly for the sake of neatness — all the way to the throttle butterfly, the entire system should be sealed to prevent air leakage and the loss of any ram gains made. If using a carbureted engine it's also critical to equalise the float bowl pressure by running tiny vacuum lines to the float bowls that run off your ram effect duct. Motorcycles have been doing this for years. Have a look at a current ZXR Kawasaki for an example of carby equalisation.

Thankfully, with fuel-injected cars it's no problem as in most cases the factory injection system can sense the extra flow of air (in mass flow systems such as the Ecotec V6 and VL Series) and adjust the fuel mixtures accordingly. With a MAP sensor-type system, as used with the VN-VS Commodores, the extra ram effect can cause a slight lean-out of air-fuel mixture once you start making positive pressure. The reason for this is the MAP sensor is only calibrated for absolute zero boost at maximum — possibly a slight chip change to enrich the top end of the fuel scale could be in order for those of us who belong to the high-speed club.

Our experiment here was to fabricate a ram air system for a V6 Commodore without spending the national debt and then see if we could measure on a dynamometer the effect of the installation. After all, if we couldn't measure the change using the ram effect from the dyno cooling fan, it probably wouldn't be worthwhile doing on a road car. We have tried very hard to use off-the-shelf parts to keep the job as cheap as possible — and also, for those of us who would rather have a trusted workshop undertake the job, it should not cause any problems or great expense.

The key in installing a decent ram air pickup on a Commodore is the type of ducting material used. You must always remember that the standard airbox on a VN Commodore onwards is quite a good thing. Look at aftermarket EFI conversions; they all seem to like using them. There are only two main problems with the Commodore airbox:

1) The standard air pickup next to the radiator does not gain any ram air effect worth mentioning. The incoming charge of air is also slightly preheated by the air thrown off the fan and radiator.

2) The standard panel insert air filter has been known to be restrictive due to its manufacture from paper/cardboard.

Good-quality ducting is available from a number of places, from irrigation suppliers to specialist race shops that sell large-diameter brake ducting. Generally, you'll find that irrigation shop material is cheaper but of a lower quality in terms of abrasion resistance and so forth than the proper race gear. Try to keep the diameter of the pipe in the region of three inches if possible. It doesn't hurt to have a larger-than-normal inlet (the standard airbox port on a Commodore has around 3-3.5 inches of cross-sectional area). Under no circumstances do we want to risk choking the engine with less airflow capability than standard.

You'll notice that the ducting we have used in this example is of the more expensive,

7

For excellent flow & filtering capabilities a K&N filter element is used.

The preoiled K&N has a million-mile warranty.

8

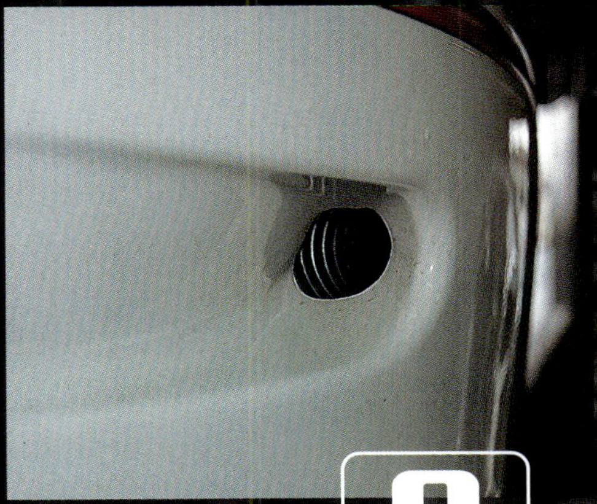

9

Standard bumper duct is a perfect place to grab ram air.

highly flexible variety with a stiffening wire insert. While costing a lot (around $30 a metre!), this is much more flexible than the solid plastic convoluted type, which is still quite acceptable but slightly harder to work to shape.

The ram air pickup location on this Commodore was super easy, with the General already supplying us an unused ready-to-go port on the front bumper. This is a near-perfect duct with a beautifully streamlined and radiused inlet to promote air grabbing. If you're not lucky enough to have a pickup point like this, going behind the grille and obscuring as little radiator area as possible seems to be the best bet.

The practice of hanging ram pickups below the level of the front spoiler to get an uninterrupted airflow is foolhardy, to say the least, increasing your chances of getting an engine full of water (with the resulting internal damage) should you ever have to drive through a puddle. If you can't safely position your duct up and out of the way, DON'T USE THIS TYPE OF AIRBOX for the sake of a few kilowatts.

It's always a good idea to use some kind of 'rock stopper' on the front of your pickup if you can, to stop the bigger pieces of road trash bombarding your air filter element.

Without some sort of protective wire gauze it's possible to end up with full-size pieces of gravel in the bottom of your box. While this doesn't actually hurt anything (it's still on the outside of the filter) it can promote filter wear (sandblasting by constant bombardment), which is why I recommend using a high-quality wire-reinforced K&N element. Not only does the K&N have better airflow properties than the factory paper unit, it's also about 20 times more resistant to damage from airbox trash.

The ram air tube is fed into the standard airbox by using a three-inch holesaw and going straight through the bottom and into the inner guard. Make sure that when the air tube runs through, it's a good tight fit. It also pays to run a sealing bead of silastic around the outside of the ram air pipe to the airbox. Remember, too, to make an aluminium or steel plate to blank off the factory forward-facing air pickup in the airbox. This plate can be simply riveted in place and then, of course, sealed with another fine bead of silastic to stop ram air leakage.

On the dyno, it was obviously very difficult to calculate the simulated roadspeed ram effect from the dyno fan. Rather than use a conventional manometer, I opted for an old aviator's airspeed gauge used back in the days of the Wright brothers to determine aircraft speed. The airspeed from the dyno fan (when located in front of the duct) resembled the airspeed figure generated at 65km/h on the road. While not giving us startling results, we should be able to measure a slight (very slight) difference on the Dyno Dynamics rollers.

The results of the dyno test were quite surprising, with the Commodore V6 showing a gain in maximum power from 94kW to 97kW, a gain of 3kW overall, but a percentage gain of 3.2%, most of which, we believe, was generated by the cooler inlet air charge than any possible boost effect. But let's face it, as long as we're making the free power, it doesn't really matter whether it's temperature or pressure that helps us produce it.

10

The flexible tubing is run under the wheel arch into a 3-inch hole holesawed in the guard.

Make sure the wheels can never hit pipe under lock to lock steering.

11

12 Remember to blank off the original air feed or you will lose effect.

13 With the airbox back in place the tube can be seen feeding into the bottom.

14 Neat, tidy and subtle, you will never know by lifting the bonnet.

If you sit down and calculate the cost of doing this type of job it's not too bad:

Flexible ducting	$40
Rock stopper	$40
K&N filter insert	$95
Cable ties & rivets	$5
Steel plate	$2
TOTAL	$182

The well-heeled owners out there can have a super professional job done by most good workshops (such as Tilbrook Auto & Dyno) for around $100 extra.

Cya on the street.

DonnonTech
267 Gorge Road, Athelstone SA 5076
Ph: 0414 437 356 Fax: (08) 8336 7886
Internet: www.senet.com.au/dtech
Email: dtech@senet.com.au

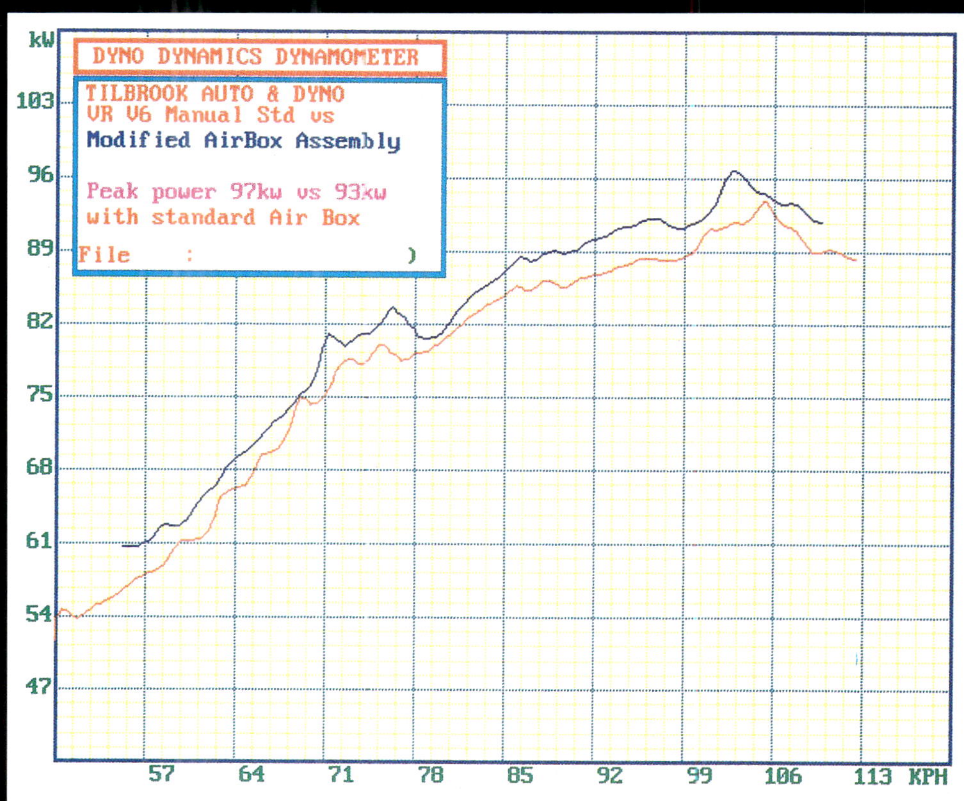

How To

MANUAL TRANSMISSION ROUND-UP

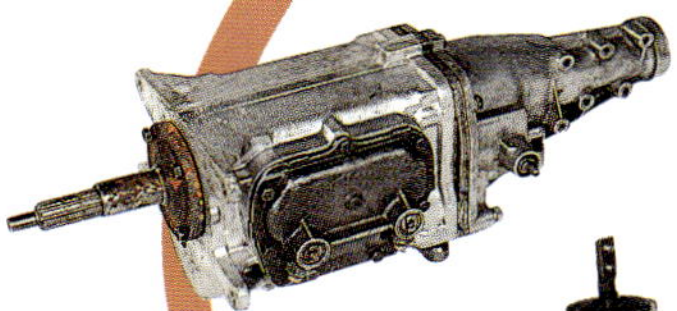

No one can deny auto transmissions are nice and easy around town, but for precise driving feel on your favourite set of curves, a good manual would have to be the enthusiast's choice. We take a good look at some of the best

There are so many different gearboxes available, sorting through them seems a daunting task for the new street machine enthusiast. We thought a general round-up of some of the more common types, considered suitable for performance V8s, would be helpful. Some of these are still readily available while others are becoming increasingly rare. This article should help you through the maze.

MUNCIE

This name has suffered all sorts of unintentional mispronunciations over the years, 'Muncie' and 'Munzie' being two of the most common. Just pronounce the c as an s and you'll be on the right track. The name comes from the American town in which General Motors manufactured these units.

The Muncie was used from about 1963 to 1974, when it was discontinued in favour of the Super T10. All versions of the Muncie had aluminium cases and although they were plentiful in the '70s they are now becoming fairly rare.

You can tell an early Muncie from a later model by the drainplugs. Early versions had only one drain plug situated high on the side of the case. Although there was provision for another at the bottom, it was not tapped out until later versions. In the early versions, a 7/8″ lay shaft was thought to be strong enough, but later this diameter was increased to 1″. Later versions also had heavier rear housings. In the general evolution of GM gearboxes, the Muncie fits between the T10 and the Super T10, having replaced the former in about 1963.

The Muncie came in four versions: a close ratio 2.2, a 2.52 version, the famous 'rock-crusher' and a 'big Muncie', which was available for about a year on HQ Monaros fitted with 350 Chev motors. This last type featured a fine 26-spline input shaft, larger diameter output shaft that fitted the Turbo 400 yoke, and stronger internals. All other versions of the Muncie had 10-spline input shafts.

On the matter of internals, everyone wants the famous 'rock-crusher' version of this box and you can certainly find people who will sell you one. The problem is, there really weren't many of these boxes around and there is no way to tell a 'rock-crusher' from a standard Muncie by external examination. About the only clue is the gears have a straighter cut for greater strength and even this is difficult to see. Beware the supposed 'rock-crusher', because it was only available as an option on very high-powered cars like the 454 Corvette. Rod Hadfield from the Castlemaine Rod Shop says there really weren't very many genuine ones out here.

Because Muncies were factory high-performance boxes they don't give much ►

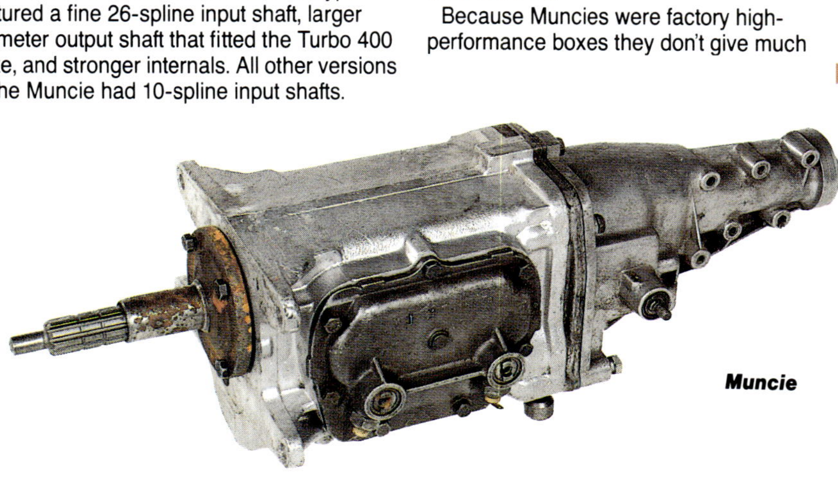

Muncie

Story and pix: PAUL TUZSON

trouble. They just wear out their replaceable parts that are still available. Rod says the amateur mechanic should stay away from these boxes because there are many things you can do incorrectly when working on them. As an example, the lay shaft is designed to be both inserted and removed from only one end of the case. This is because one end of the shaft has a very fine 0.0001-inch taper. The case is aluminium so the shaft can be bashed out the wrong end. When the new one is fitted, though, one end will be slightly loose because of the stretched hole in the case. This is just one of many potential problems.

Saginaw

SAGINAW

Like the Muncie, the Saginaw is also named after the town in which it was manufactured. It, too, suffers from mispronunciation, having been called 'Saganaught', 'Segenor' and all sorts of other things. The correct pronunciation is a combination of sag, as in what old things do, and inaw(e), as in your state of mind the first time you see a Top-Fueller spit down the quarter-mile!

The Saginaw was a pretty good four-speed found behind Holdens from 1968 to 1970. It was available behind 253s and 308s as a special order. Holden listed it a heavy-duty four-speed but most experts agree it really wasn't all that heavy duty, although all Saginaws had iron cases.

Many gearboxes have annular grooves machined around the splines on the input shaft. On any particular box, these indicate the first-gear ratio. Rod has been working on Saginaws for years and his notes indicate the following: no grooves equals a ratio of 2.85, one groove means 2.54, two grooves represent 3.11, and three grooves indicate 3.5.

American versions of this box had the shifter bolted to the tail-housing. In Australia the shifter was bolted to the floor, so our tail-housings had no provision for the shifter. This is how you can differentiate between American and Australian versions.

Saginaws had the same total case length as Muncies and the same bolt patterns as Muncies and Super T10s. They stayed much the same throughout the period of their availability and only came with the same 10-spline input shaft as the Muncies. Spare parts are rare and, although new parts are available, they are expensive. This box was prone to mechanical problems and was disadvantaged by shifting links that came out of adjustment.

People have confused Saginaws with Muncies. To identify them, just remember Saginaws have their reverse selectors mounted on the side cover while those on Muncies are on the rear extension housings. Additionally, all Saginaws have cast iron cases while, as we mentioned previously, all Muncies are aluminium. If the box isn't mounted, lifting it should tell you what it is. It is interesting to note in comparing Saginaws with Muncies, the heavier box is the weaker.

SUPER T10

The Super T10 is by Borg Warner and comes in four ratios from close to ultra low. Annular grooves indicate the ratios; the more grooves, the lower the ratio. It is still manufactured and enjoys great popularity among Commodore racing people. In fact, the Super T10 was developed for racing. Its forebear (simply called the T10) was introduced in the late '50s and early '60s behind Corvettes.

Features of this high-performance box are: a fine (26-spline) input shaft; double-row needle-roller bearings at either end of the main cluster; larger diameter mainshaft; gears designed for better lubrication and cooling; bigger bore gears; a 32-spline output shaft which lines up with a Turbo 400 yoke; and a choice of aluminium or cast iron cases. This is a very strong gearbox. It has a dedicated following which argues it is as strong as the legendary Top Loader from Ford. Having witnessed one of these arguments I'd hate to offer an opinion.

In restricted classes of racing, people have been known to take a T10 case and fill it with Super T10 internals. Because the acceptability of a gearbox is determined by an external examination, competitors are able to make use of a much stronger gearbox that would otherwise be banned.

TOP LOADER

This would have to be one of the most well-known gearboxes around. Everyone has heard of its legendary toughness. Although the reputation is well founded, I have still found people who have broken them. I guess nothing is immune to abuse. Rod Hadfield says when they break it is often at the circlip groove for the speedo drive.

They came in wide- and close-ratio versions and had 1 1/16″ 10-spline input shafts. There was a bullnose version that also had a 10-spline input shaft but with a larger diameter of 1 3/8″, however, this was fairly rare and is now quite expensive. It also had a 31-spline output that was the same as an FMX auto. This was the version fitted to the equally legendary Phase III HO.

The shifters on these boxes were attached in two ways. Until the XA, they were bolted to the floor. Because the floor and the gearbox moved independently of each other, these versions could jump out of gear. After the XA, the shifter was bolted to the box itself, overcoming the problem. Adaptors are available to fit this gearbox to almost anything. If you're looking for a Top Loader, the best ones to get are those that came out of six-cylinder TC and TD Cortinas because they won't have had so much power put through them. Parts are becoming harder to find but Rod says they're around if you look hard enough. All versions of the Top Loader

Super T10

had cast iron cases. Currently, there are after-market replacement nose cones available in anodised aluminium.

Top Loader

BORG WARNER SINGLE RAIL

The Single Rail was found on V8 and six-cylinder Falcons from XB onwards, V8 and six-cylinder Chryslers and Cortinas, and V8 Leylands. Some Datsun 200Bs and some Sigmas used a lighter version of the Single Rail, and even Holden used a light version for a while. Ford V8s were fitted with two versions of this box: The 302 version had a lower ratio first gear than the 351 due to the latter motor's greater torque.

The Single Rail uses the same spigot bearing as the Top Loader and when mounted, one bolt hole matches. Fitting it requires the remaining three be drilled and tapped. Although they're not as strong as Top Loaders, Single Rails can be toughened up somewhat by fitting new bearings front and rear. Again, adaptors are available to fit this box to practically any motor.

Borg Warner T5

Buyers should beware of a couple of potential problems in trying to fit a Single Rail. It is a good idea to try to buy the correct tailshaft with the box, particularly if it's from a Leyland or a Chrysler, because no other yokes fit these boxes.

Additionally, you can't put a Single Rail from some other car into a Chrysler because they need to have a specially bent selector rail. In fact, you can always tell a Chrysler unit by the U-bend in the selector rail that is there to clear the floor member. Fords are identifiable by their shorter-than-normal selector rail. Parts for the Single Rail are no problem and overhaul kits are still available. Speedo cables between the various boxes are not interchangeable but there are adaptor kits available.

BORG WARNER T5

In Australia, the Brock Commodores had the first T5s and these had fine (26-spline) input shafts. There is, in fact, a large variety of input shafts available for these boxes. This is a current box and is constantly being upgraded. As such, it has very fine tolerances and parts from one box shouldn't really be used as replacements in another because the wear patterns should match. If they don't, I'm told a noisy box will almost certainly be the result. Because of the fine tolerances, it is vital to use the correct oil in these boxes.

There are no aftermarket parts made for the T5. Everything comes through dealers carrying the sort of price you'd expect from such an arrangement. Again, when fitting a secondhand box, it is important to know where it came from, for conversion purposes, but is often difficult to identify the various versions.

In Camaros, the T5 was mounted at a severe axial angle to bring the box-mounted shifter over close to the driver. In Australia, we sit on the other side (had you noticed?) so the boxes had to be mounted upright. This entailed setting the cross-member at an angle to compensate.

Single Rail

TREMEC TR-3550

When the Top Loader was discontinued because of concerns over high-powered cars and pollution in America during the 1970s, all enthusiasts were sad to see it go. Fortunately, the Tremec company in Mexico was able to do something about it. It bought all the tooling from Ford and continued to make a version of it, in fact, Ford continued to buy the Tremec 'Top Loader' for certain vehicles.

As time passed, the Tremec version evolved into the alloy case, five-speed unit shown in the photograph. The addition of the fifth speed, which is an overdrive, is a real advantage in a performance car fitted with a low-ratio differential. The fifth gear is in the rear extension housing.

This gearbox is available in three versions (Standard, HP and TKO) each of which handles greater power and torque. The price also rises with the performance capabilities, but Tremecs are not as expensive as other brand new high-performance boxes. You also need to consider the other advantages. Being brand new, the boxes are covered by a warranty, and all parts are readily available. In addition, they have the same bolt pattern as a Top Loader and the shift lever can be located at three positions along the top of the box.

The photograph shows it in the rearmost position with the other two mounting points

Tremec TR-3350

covered by plates. Tremecs are one of those products that allow you to mate 1990s' technology with an older car. In America, the Ford Special Vehicle Team has chosen the Tremec for use in their special-build Cobra-R model Mustangs.

The boxes we've mentioned here are certainly not the only options, they're just the ones most people relate to performance V8s. There are others more expensive and others that are cheaper. Of the latter type, secondhand boxes out of Celicas and Supras can be a worthwhile investment, particularly if you have a four or a six. Such boxes are also regularly used behind V8s.

Bob Coleman, who works as a consultant to Tri Automotive, gave us some of the information for this article, while Rod Hadfield from Castlemaine Rod Shop was also very helpful. Both businesses carry parts and full transmissions. Tri Automotive has a free-call info line on 1 800 815 813 and the Castlemaine Rod Shop is on (054) 722 853.

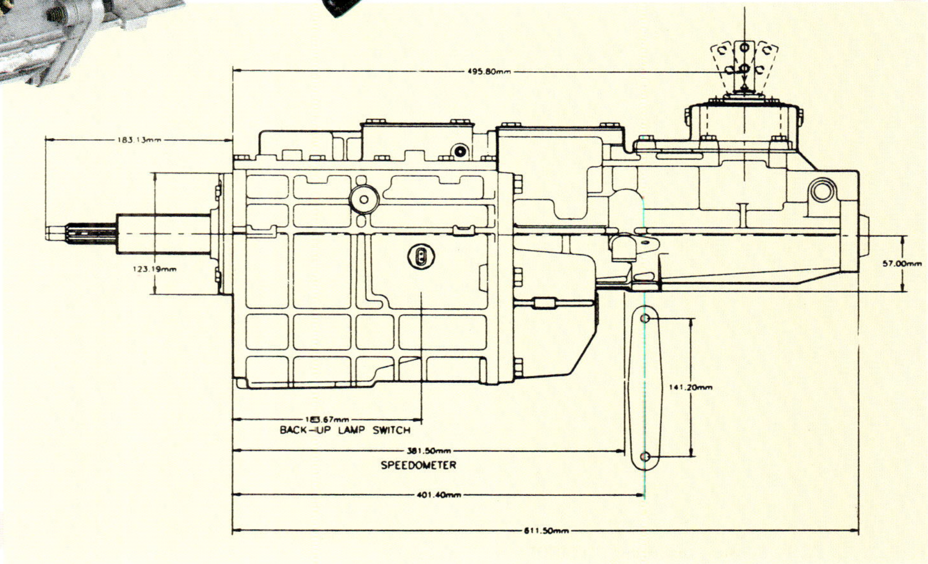

REBUILDING THE BORG-WARNER SINGLE-RAIL

The single-rail is not renowned for being a particularly high-performance gearbox yet Howard, our gearbox specialist, says they really are a very good box. They're not as robust as top-loaders but they're still good strong boxes that came in a number of ratios for Cortinas, six-cylinder commercial vehicles and Ford V8s. Howard calls them Borg-Warner's finest. He says, in fact, that in medium-power applications, they don't have any problems at all. There are still plenty of them around, so we thought we'd show how to fix and reassemble one.

Our usual warning still stands: don't attempt what we're showing you here unless you have a factory manual. Read this article and relevant section in the manual before you do anything. This is designed to be a complement to the manual, not a complete guide in itself. Read the article backwards for clues about dismantling and, of course, your manual will tell you how to go about this aspect of the job. As you disassemble the parts, lay them out in positional order on sheets of butcher's paper. Make notes on the paper as you go.

1

2

3

4

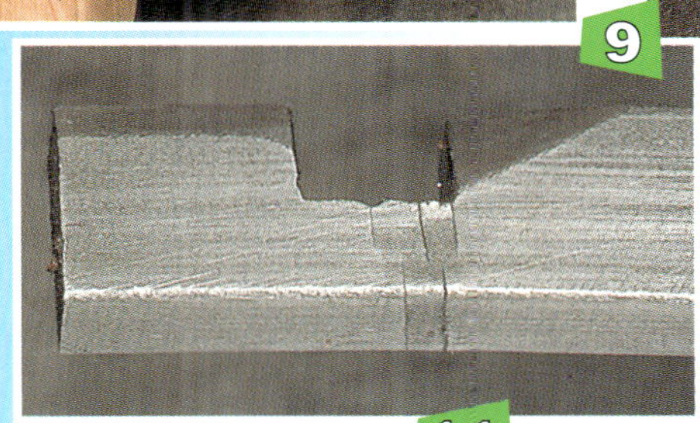

1. *After disassembly, you'll need to check certain parts for wear. These are two selector interlocking collars. The one on the right is worn and must be replaced with the new example on the left. They're made from special material and can't be repaired by welding.*

2. *As an example of the in-built strength we mentioned, this shot shows a lay-shaft from a single-rail compared with that of a Holden four-speed. No contest.*

3. *One problem that occurs with the blocking-rings is they get knocked back in the areas shown. The example on the left is damaged while the one at right is new. If yours are showing signs of this effect, they must be replaced.*

4. *The other thing to watch with the blocking-rings is that the internally machined annular grooves stay sharp. This is an area that doesn't normally wear too badly. Run your finger across the grooves and see how much bite they have.*

5. *This shim controls the mesh of the input-shaft and lay-cluster. It must be placed on the input-shaft as shown before the bearing is pressed on. Other makes of gearbox have similar shims but they aren't always as important as this one.*

6. *Fitting the synchro-keys to the inner synchro-hub and then slipping the outer hub over the assembly as shown is fiddly — enjoy! You can actually fit the inner and outer hubs together the wrong way around, so don't. Make sure they're facing the way Howard is holding them in the shot.*

7. *The synchro assembly is held together with these clips that lock into the synchro-keys as shown. There are two clips and one fits on either side of the synchro-hub assembly. Make sure the openings on the clips face different ways, as shown. If the openings are aligned with each other, you'll have problems. Each clip should engage all three pins.*

8. *This shot is just making sure you get the inner and outer hubs assembled properly. The ridge Howard is pointing to should face the way it does in the photograph in relation to the outer hub.*

9. *Another problem you may find after you've cleaned all your parts is the selector forks can crack at this point. These can be repaired by welding, but after the job is done, these sections of the selector forks must be ground and filed back to their original shape.*

10. *One very common problem with single-rails is that the threads on the selector lever assembly and housing get stripped.*

11. *If the male threads on the nylon shifter nut are stripped, fitting the replacement unit involves disassembling the lever assembly. Don't try to do this by removing the top part of the lever with the metal sleeve. It's a factory fit and if you do get it off, you won't be able to get it back together again. Rather, remove the circlip Howard is pointing to in the next photograph, then pull the parts up the lever as shown. This will allow access to the ball, which should be cut diagonally as shown. You'll then be able to remove the ball and slip the damaged nylon selector nut off the bottom of the lever.*

12. *After you've fitted the new part, replace the cut ball and slide all the components back down the lever into their original positions. Compress the spring and replace the circlip. If the female threads are the damaged ones, you can have the threads re-cut oversize and fit an oversize nylon shifter nut.* ►

REBUILDING THE BORG-WARNER SINGLE-RAIL

13. Having carried out all the repairs and checks mentioned so far, it's time to start reassembling everything. Clamp the output shaft in a vice with soft jaws as shown and set second gear and its blocking-ring in place, as shown.

14 & 15. Next, the first/second synchro hub goes on. This must be firmly seated, so it should be tapped down using a drift as shown. A good place to go and get odd bits of pipe and whatever is the scrap bin at your local steel supplier.

16. The synchro hub is held in place with a circlip — don't forget it.

17 & 18. Set the first gear blocking-ring in place, followed by first gear.

19. Now it's time to fit your replacement rear bearing. This must be pressed in place — there's no other way — so you'll have to get access to a press. Make sure the circlip groove faces rearwards.

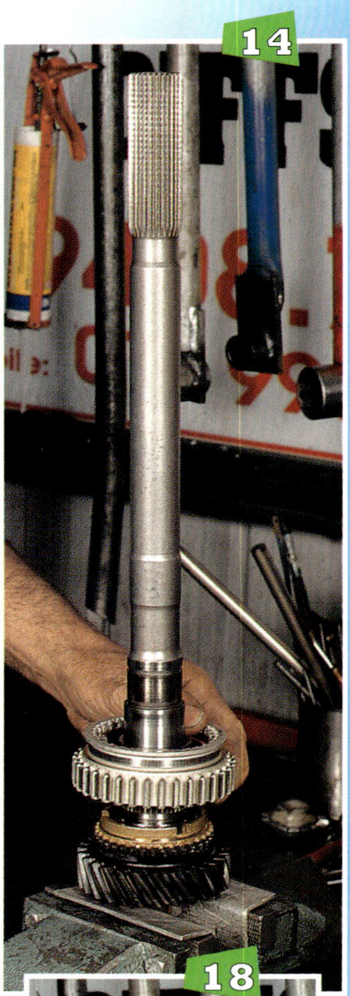

REBUILDING THE BORG-WARNER SINGLE-RAIL

20

21

22

23

24

20, 21 & 22. *Next, while the assembly is still in the press, third gear and its blocking-ring is added, followed by the third/fourth synchro hub. Don't forget the circlip.*
23. *Fit the reverse idler gear as shown. Make sure the pin on the selector fork is engaged in the groove.*
24. *There are two rows of rollers on each end of the lay-shaft. Obviously, they're not applied in this manner. We just stuck them on the outside of the shaft so you could see how everything sits. The way to do it for real is to stand the lay-cluster on its end, drop the first of the locating rings down the hole and position the rollers around the bore. When the inner row is done, another ring is dropped down the bore and the second row of rollers goes in. These are retained with the thickest of the three locating rings. Repeat this procedure for the other end. Howard prefers Vaseline for holding the rollers in place until the lay-shaft is inserted.*
25. *The seals in the front and rear housings must be replaced.*

25

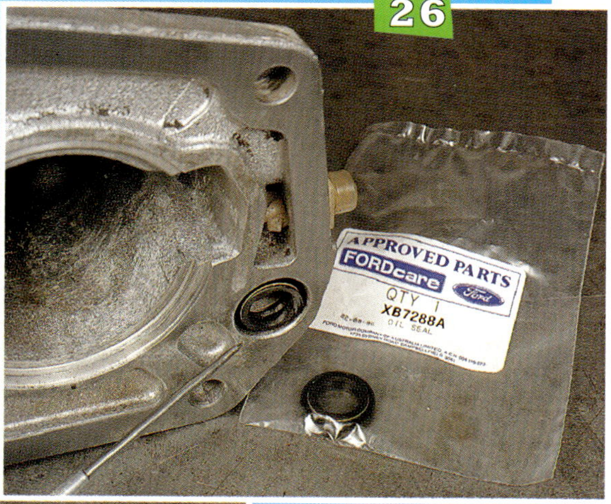

26

27

26. *There's also a seal for the selector rail. It's in the rear housing where Howard is indicating.*
27. *The lay-cluster has three thrust washers that must be replaced. The two really strangely shaped ones go on the rear. Again, Howard uses Vaseline to hold them in position during assembly. The copper-coloured sides must be in face-to-face contact. The third washer sits at the front of the cluster with the locating tabs facing out and locking into the case.*
28. *When the washer's in place, lower the lay-cluster into the case as shown.*
29. *Load the main-shaft assembly rear first as shown.*
30. *When the main-shaft is in the case, the assembly is pulled backward to expose the groove on the bearing into which the circlip is fitted. With the circlip in place, the shaft is pushed forward again.*

28

29

30

31

32

33

34

35

31. *A new bearing is pressed onto the input-shaft, which is fitted with a synchro-ring and circlip and then fitted as shown.*

32. *Tapping with a soft hammer helps seat the bearing into the case properly.*

33. *Next, the case turned upside down on the bench. Doing so causes the lay-cluster to fall into mesh with the gears on the main-shaft. Howard is aligning the thrust washers with the holes so the shaft can be inserted.*

34. *The lay-shaft is inserted.*

35. *A locking tab engages slots in the reverse idler-gear shaft and the lay-shaft as shown. A dab of Loctite secures the tab.*

36. *Fit the front and rear selector forks as shown then push the selector rail through the rear fork, selector interlock collar and then the front fork. Howard's thumb is resting on the drilling, which takes the selector-shaft detent ball and spring. The spring can be seen in the next shot. The ball engages the grooves on the shaft.*

37. *The selector fork interlocking pin must be driven in as shown. If you drive this in too far, the selector interlock collar won't slide and the selector mechanism won't work. If you do tap it a bit too far, just turn the shaft over and tap it back a bit.*

38. *Fit a new gasket and attach the rear housing. At this point, the end-float should be checked. The figure you're after should be about 0.012-0.015", which is just under half a millimetre.*

39. *If your end-float is too great, re-attach the rear housing without the gasket. If this fixes the problem, seal the housing with Silastic. If this still doesn't fix the problem (it probably will), remove the rear housing and make a shim like this that fits into the rear housing and sits against the outer race of the rear bearing. Shim material is available in small sheets from Repco.*

36

REBUILDING THE BORG-WARNER SINGLE-RAIL

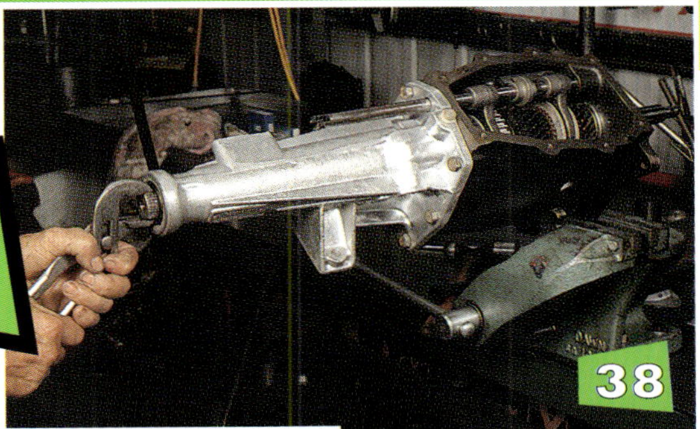

38

37

39

When the end-float is correct, refit the selector housing and the job is finished. Howard's business is Trade Trans-N-Diffs. He has all the parts you'll need for the job and you can call him on (03) 9408 7597 or 018 998 910. Many of these parts are still available over the counter from Ford.

REBUILDING A TURBO 350

A tough job but someone has to do it...

Story/photos by Paul Tuzson

Rebuilding a transmission is a tricky business, as everyone knows. Recently, however, we spoke to the guys down at Pro Automatics and asked if it could be done by a competent home mechanic. Happily, they said yes and agreed to show us how.

We're not going to talk about getting the transmission out of the car — if you have trouble working that out, you shouldn't even try pulling it apart. Anyway, we just don't have the space. As it is, there are nearly 50 pictures just covering assembly. For the same reason, there are no shots of disassembly. For clues about how that part of the job should proceed, just read the story backwards!

1. All the various parts are laid out on a table. The one important thing about disassembly is that you do not do it this way. Obviously, Pro Automatics has been doing it for years and know where everything goes. You don't. Having watched the whole job, we'd say the most important part of disassembly would be laying the parts out. We suggest you go down to the local fish shop or butcher and get a whole lot of white paper sheets. Lay these out in a long line and as you disassemble each component place it clearly in a linear sequence. Also make notes and drawings on the paper to assist in remembering where everything fits. Automatics are complicated and you'll be very glad you did! Additionally, you'll need an owner's manual. Cleanliness is even more important than laying the parts out, because even if you manage to get it back together properly, any grit in the unit will mean you'll have problems very soon after it goes back into service. As you dismantle the unit, clean each part thoroughly in either Mobil B1 or Shell X55 solvent. Don't use petrol or kerosene.

2. The rebuild kit contains all the seals and gaskets you'll need. All the items in the kit must be replaced. The kit itself will serve as a guide for finding parts during dismantling. You must find a match for all of the seals and gaskets in the kit. As you do locate them, don't just throw them away; place the used items on the paper next to the components from which they were removed. Take careful note of the direction in which components were fitted. Putting some seals in backwards will mean instant failure in service. The bushes shown in the shot should also be replaced. This is not always necessary but your problem will be judging the condition of the original units accurately. It's just safer to replace them all. Tony says you can get the bushes in and out with some care. He says screwdrivers and pin punches will do to remove them and sockets can be used to get the new ones in. He's not saying this is recommended practice, only that things can be done this way if you're attempting the job at home without a press. There are nine bushes to find.

3. Tony is slipping on a new seal. The components in each kit must be examined carefully. In fact, one of the seals in this kit had a tear. Tony says this can happen from time to time.

4. Tony is pointing to the internal seal in the direct clutch drum, one of those needing to be replaced in the correct direction. Putting it in backwards will mean slippage in reverse.

5. All the seals have been fitted to the pistons. Do them one by one and replace them on your sheets as you do them.

6. The thrust washer on the forward planetary set must be checked. If it's scored it must be replaced. Tony says about 30 per cent are bad. These are available as part of a kit or individually.

7. You must check the forward planetary set for wobble and end play. If either exists, replacement is the only option.

8. There are seven thrust washers and each must be checked for scoring. If you find any, the faulty items must be replaced.

9. The bearing surface for the pump reactor sealing rings must be checked. Tony says that to the untrained eye the score marks can look like they have been purposely machined. The surface Tony's pointing at must be perfectly smooth.

10. The seals on this shaft must be replaced with those in the kit. The originals are metal but Tony says the Teflon replacement units seal better.

11. To ensure that they spring out in the grooves Tony gives them a bit of a squash before fitting them.

12. When all seals and thrust washers have been replaced it's time to start fitting the major components together. Coating the parts with transmission fluid makes things easier.

13. Make sure your vice is fitted with soft jaws of some type. When slipping in the pistons you'll have to be careful to avoid catching the seals. Use a thin but fairly rigid feeler gauge like a shoehorn to help ease them into position. Run the gauge right around the seals. When you are inserting the pistons you should apply only the gentlest of pressure. If a piston binds during insertion something is wrong. Take it out and try again.

14. Insert the forward clutch return spring pack.

15. You'll need some way of compressing the spring pack evenly so you can fit the retaining circlip. This is a special fixture but you might be able to improvise with a drill press. If you don't have a drill press you'll have to devise some other means. Whatever you do, do it carefully. Make sure you don't slip and damage things.

16. Replace the pump seal. Tony uses the large round metal disc shown and you can improvise with something similar.

17. The gears from the pump must be inspected for any

obvious signs of wear and replaced if any is found.

18. Check the clearance on the pump gears. If it's greater than .003 you'll have to have the face of the pump machined or replace the parts with new items.

19. Reassemble the pump gears, cover, intermediate clutch piston, springs, retainer plate and bolts. Fit the gasket from the kit and this section is ready for fitting later on.

20. A band like this is used to ensure that both halves of the pump are properly aligned.

21. Replace the direct and forward clutch packs. You'll need new friction plates, which should be soaked in transmission fluid for a few hours before they're fitted. The intermediate steel clutchplates can be reused if they're not scored or burned. If they are, you'll need to fit new or reconditioned replacements. Don't forget to drop the large wave washer in place before inserting the clutchplates.

22. When the clutch packs are fitted, insert the pressure plate and check the clearance between it and the clutch pack.

23. The clearance between the pressure plate and the clutch pack should not be greater than .025. If it is, you'll need a new pressure plate. These are available in different thicknesses to achieve the correct clearance.

24. Fit together the forward and direct clutch assemblies.

25. Fit the output shaft/rear annulus into the case. You should check the shaft for scoring before refitting.

26. Insert the reaction carrier.

27. Reassemble the low/reverse clutch with new friction plates, not forgetting the spring washer first.

28. Insert the low/reverse sprag clutch assembly. Tony shows the centre support reaction spring and the position in which it engages the centre support. The spring is actually mounted into the case and the centre support mates with it during assembly. During engagement of first from neutral this spring takes up the shock as the sprag clutch locks up.

29. Fit the sprag bearing centre race.

30. Insert the sun gear and drive shell.

31. Install the forward planetary set. We haven't shown it, but once the planetary set is in place it's held there with a circlip. Don't forget to refit this clip or you'll have serious problems.

32. Fit the input ring gear over the forward planetary set (already in place in this shot) and then fit the forward and direct clutch assemblies.

33. Slip the intermediate overrun brake band into position over the clutch assemblies.

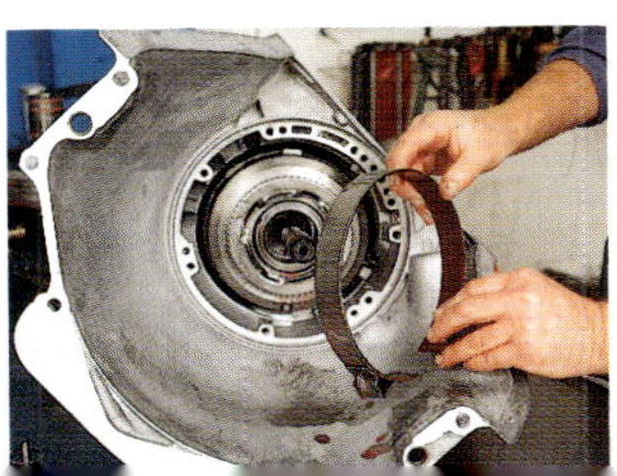

34. Load the intermediate clutch assembly. Some have only two friction plates but three are needed.

35. The pump assembly is fitted with spacers which govern the clearance. Mount a couple of them on the shaft.

36. Guide the assembly into place using a screwdriver to match the holes indicated. There's only one way in which you can fit the pump unit.

37. With the unit properly seated, check the endplay. It should be no more than .020-.045. If it's not within these figures, add or remove washers as needed and tighten to specifications.

38. Insert the piston for the intermediate overrun band.

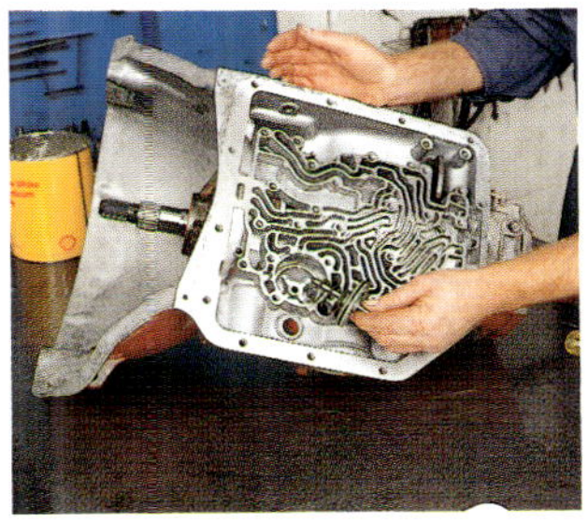

39. There are four ball check valves in the valvebody. If you'd like to modify your transmission for a quicker shift between all ratios, leave them out except for the one that goes in the position Tony indicates.

40. A further performance enhancement can be had by drilling out these two holes from the original size to $^{3}/_{16}$". Tony holds a modified plate over an unmodified original so you can see the difference.

41. Mount the gasket and separator plate. In a full shift kit the separator plate has a slot milled in the upper face to allow faster fluid transfer to operate the top/reverse clutch. This one doesn't. We're just showing you a few tricks that won't cost you anything.

42. Fit the valvebody back into place and tighten the bolts to specification. When you dismantle the valvebody to inspect the components it contains for wear, remove the valve assemblies only one at a time. Replace each one before moving on to the next. That way you won't get confused by having lots of different parts rolling around everywhere.

43. Refit the one/two accumulator piston.

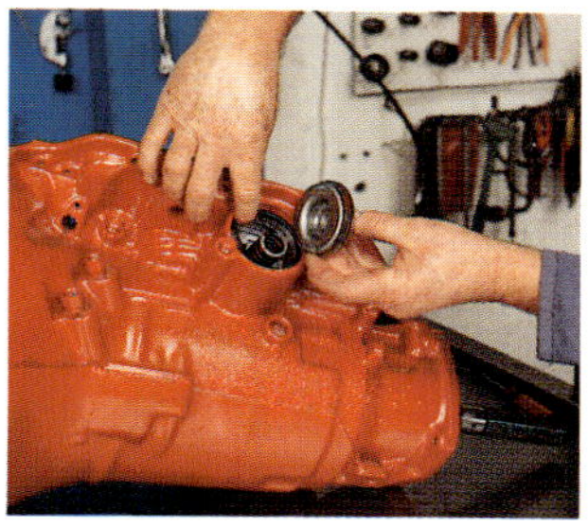

44. Replace the governor. This is another component that can be altered free of charge. Lightening the centrifugal weights will cause a later change. Try filing about a couple of millimetres off the end of each weight.

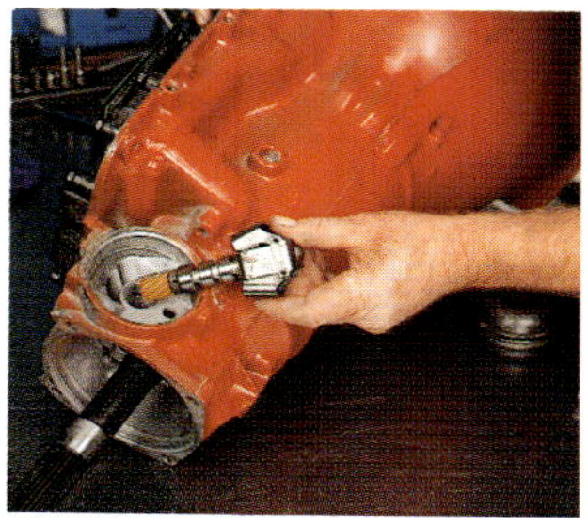

45. Refit the gasket and pan.

46. Fit an adjustable modulator. Turn the screw clockwise for later, firmer shifts.

47. When fitting a Chev housing to a 308, many people use adaptor plates. Tony doesn't like the practice because a spacer must be used to bring the torque converter back to the input shaft. Once the converter is moved off the flexplate in this way, vibrations tend to occur and cause pump and other problems. Rather than a spacer plate, Pro Automatics has housings weld-modified to take the correct bolt pattern. The unit we got them to assemble for us will have to be pulled apart again because some extra holes have yet to be drilled.

This has been a general overview of what's involved in rebuilding a transmission. We've by no means covered every detail and if you do decide to attempt the job you really should have a factory manual. If it doesn't look too daunting give it a go. But please, use the white-paper method for dismantling — it's the only way you'll get everything back together. Tony says the whole job would take him three and a half to four hours, including dismantling and cleaning once the transmission is out of the car. An amateur would probably need to set aside a full weekend for the job and, even then, you probably wouldn't want to make plans for the evenings.

Thanks to Ron and Tony at Pro Automatics for their help with this article. You can reach them on (03) 9791 5427.

TOUGHENING UP THE BORG-WARNER REAR END

For years a motor with a decent power rating has meant having a Ford nine-inch rear to split the torque. No more!

There's nothing wrong with a nine-inch; in fact, I have one. But it's common knowledge that the weight and large gears of Ford's monster rear-end will sap substantial power from a drivetrain. Still, there hasn't been a reasonably priced alternative; if you had more than about 350 horsepower you just had to get the conversion done or keep breaking your diff.

Vas, from Keysborough Diffs, has been doing such nine-inch conversions for years, but now he's decided to do things a bit differently. Why bolt all that extra weight under your car when you don't really need to?

On the surface, OE gears and the new types look fairly similar.

Many aspects of the Borg-Warner differential will handle quite a bit of power. It's really only a matter of replacing the components that aren't up to scratch. Unfortunately, two of the main weak points in a Borg-Warner are the ring and pinion gears. They're just not cut out for major stress and, unfortunately, designing and manufacturing replacement sets is a fairly costly exercise. Other common problem areas are excessive wear in the lsd clutch cones, fatigue and breakage of the lsd triple springs that sit in the middle of the spider pinions, and some inherent weaknesses in certain housings.

NEW TEETH FOR ENGINES THAT BITE

Vas has now started manufacturing replacement parts that fix most of these problems. Let's start with the ring-gear and pinion.

On the surface, OE gears and the new types don't look much

An original equipment ring-gear and pinion.

Replacement parts (top) have greater curvature of the teeth.

Upgraded replacement parts from Keysborough Diffs.

different (Photograph 1). But look a little closer (Photographs 2 & 3). Photograph 2 shows an original equipment ring-gear and pinion while Photograph 3 shows the upgraded replacement parts.

There are two main differences between an OE ring-gear and one from Keysborough Diffs. Photograph 4 shows the first of them. Note the curvature of the teeth. The OE version is on the bottom while the performance replacement is sitting on top. The teeth on the replacement unit have a more pronounced curvature that effectively increases the surface area through which driving force is transmitted and reduces stress. The extra curvature also braces the tooth more firmly and means there's more metal at the base of the tooth that needs to crack before the tooth will break.

The angles of the upper edges of the teeth on each gear are different and this reduces stress.

Longer curves are not the whole solution to the problem. Photograph 5 shows both gears edge-on. The OE unit is on the left and, clearly, the angles of the upper edges of the teeth on each gear are very different. This has the effect of redirecting some of the forces at work in the gears. As the teeth on the pinion press on the teeth of the ring-gear, Mr Newton's third law tells us the teeth on the ring-gear press back against the pinion with equal force.

This reactionary force applies side pressure to the pinion shaft and is responsible for broken teeth. Increasing the engagement angle of the teeth to that shown at right converts some of this side-force into forward force that acts linearly against the rear pinion support bearing instead of trying to bend the shaft sideways. Changing the curvature and engagement pattern of the teeth also has the effect of lengthening the teeth on the pinion gear. As with the ring-gear, this means the stress is spread over a greater area. In Photograph 6, the performance pinion is on the right.

TOUGHER SPRINGS & OTHER THINGS

The triple spring pack that causes the engagement of the lsd clutch cones is a bit of a problem in standard units. The springs tend to be too soft and lose their elasticity after a while due to fatigue. This results in them holding the cones with a breakaway force of only about 45-50ft/lb. Keysborough Diffs supplies replacement units that increase the holding force to something in the order of 120ft/lb. Photographs 7 & 8 show the new springs, which don't really look any different from OE units.

The next area that needs attention in a Borg-Warner is the lsd clutch cones themselves (Photograph 9). OE versions wear quite severely (Photograph 10). The pair on the right are OE versions. As such wear increases, the cones move further out away from the centre, causing the wear shown in Photographs 11 & 12 as the sides of the cones start to rub in places where they shouldn't be making contact.

Vas did some investigating and found that 4140 steel is much harder than the metal from which the OE versions are made, so he's had replacement units made from this material. But simply being harder is not enough. Materials that are going to be in rubbing contact need to be compatible and 4140 steel and the iron from which the centre is made work together very nicely under these hot, demanding conditions. It took quite a bit of research to establish this.

LONGER SPLINES

Rubbing wear is not the only problem with OE cones. Photograph 13 shows the outer faces of the performance cones on the left and OE versions at right. Note the performance units have a flat surface while the others are shaped with a groove around the edge. The OE versions are thinner, which means the splines for the axle are shorter, 7.5 millimetres to be exact. By making the outer surface flush in the performance replacements, the hole through the centre becomes deeper, which means the axle splines in the new units can be 12 millimetres long.

The new units shown hadn't been broached yet, but when they are it will be much more difficult to strip them. Keysborough Diffs will ➤

The performance pinion at right has longer teeth.

TOUGHENING UP THE BORG-WARNER REAR END

Standard springs are too soft and lose their elasticity after a while due to fatigue. New units increase breakaway force to 120ft/lb.

Borg-Warner lsd clutch cones wear quite severely.

The increased wear can be seen in the pair at right.

broach these out to whatever number of splines you require. Fitting these thicker cones means the housing (Photograph 11) has to be machined down inside. The ridge down in the centre with the semi-circular notch in Photograph 11 is removed, plus another 1.5 millimetres. Vas says this doesn't affect the strength of the unit.

The last thing to consider is the housing itself. Standard units come in two versions and the earlier units are the ones to get. Later units suffer from being a bit weak at the points where the extension tubes are welded into the cast centre. The earlier cases (Photograph 14) are the ones to get.

WHAT ABOUT THE OUTER ENDS?

Having toughened up your centre, you might want to consider the outer ends of your differential/axle assembly. One of the problems facing enhanced power ratings is getting all the power to the ground without wheelspin. Wider wheels and tyres are the answer, but these can often stress normal axles beyond their original design tolerances. Because normal axles are joined to the wheel mounting flanges, a broken axle means the wheel comes off. Obviously, this is not desirable. The answer is a fully floating hub, which Keysborough Diffs has started to manufacture.

In a floating hub system, the axles and wheel mounts are not one piece as with a conventional setup. Instead, the axles are splined at both ends and the wheels turn on special hubs that are securely joined to the axle housing independently of the axles. The original ends of the housing are removed and special tubular centres are welded in place, as shown in

Replacement units are harder 4140 steel, which is totally compatible with the housings.

The cones wear on both the inner and outer surfaces.

The housing itself can be a weak point. Earlier versions are tougher.

15 *The axle housing extension for a fully floating hub.*

16 *The mounting flanges for the bolt-up faces of the wheels are made in two parts.*

17 *Two bearings are used in each hub.*

TOUGHENING UP THE BORG-WARNER REAR END

Photograph 15. The axles pass through the centre of these new extensions and they also serve as mounting points for the centre races of the new hub bearings.

TWO BEARINGS

The mounting flanges for the bolt-up faces of the wheels are made in two parts, shown in Photograph 16. The longer part of the mounting flange shown (between the axle housing and outer flange) contains two bearings that allow it to rotate on the new axle-housing extension.

Photograph 17 shows two bearing sections of the new hubs with the bearings that will be pressed into either end. When the bearings have been inserted, each inner section of the hubs is mounted onto its axle-housing extension, as shown in Photograph 18. A threaded ring holds the bearing section of the hubs onto the extension. You can see how the axle protrudes through this section of the hub.

Next, the driven section of the hub is fitted onto the end of the axle-shaft (Photograph 19). This is a dummy unit, so the hole in the centre of it hasn't been broached to fit the splines on any particular axle-shaft. Nor have the wheel-mounting studs been fitted to the double flange assembly of the hub. However, that's the whole point with this system — it's completely customisable. No matter what the

18

TOUGHENING UP THE BORG-WARNER REAR END

Both sections of the hub on the housing. The flange can be drilled to suit any application.

bolt pattern for your wheels may be, or how many splines you have on your axles, this system can be adapted to fit your vehicle. You'll be safe in the knowledge that if your axle does break under power or in a tight corner, one of your wheels won't come off and over-take you.

Having a full floating hub system like this means your engineer will be happy to certify your car with quite wide wheels and the authorities will also be happy. You can call Vas at Keysborough Diffs on (03) 9706 3454 for more details. The prices for these fully floating conversions can be found in the Keysborough Diffs ad in this issue. Vas will also do all normal conversion work on any type of differential (Photograph 20).

TOUGHENING UP THE FORD NINE-INCH

Words and photographs by Paul Tuzson

Three mini spools and one full spool.

So far, we've shown you how nine-inch conversions are done and how to assemble a single-wheeler and get the engagement pattern right. This time around, we thought we'd show various performance upgrades for the nine-inch. As usual, we went to Andrew at Supa Trik for the story.

As everyone knows, the Ford nine-inch is really tough. However, nothing is indestructible and when you start to pump enough power through almost any component you'll be able to break it. OE factory parts are most often made to tighter budgets than specialised aftermarket performance products, so it's not too long before you'll reach the limitations of OE components.

GETTING LOCKED UP

One of the first things you need to do when you get on the quarter-mile with some serious power is lock both wheels together so they turn at the same speed and both work to provide grip. If one wheel spins you'll be wasting power. Not only that, if one wheel starts spinning faster than the other, the effect will be to steer you into the wall. Most people know that fitting a mini spool or a full spool is the answer, but Andrew says he gets a surprising number of people ringing who don't know the difference between the two.

In **photograph 1** there are three mini spools and one full spool. Mini spools are smaller, simpler and therefore cheaper. They fit inside your existing hemisphere and replace the original pinion gears (sometimes called spider gears) and side gears. They are splined on both sides and your axles slip into these splines instead of the original side gears. Mini spools are mounted on the original pinion gear shafts (commonly called 'pins') so you need to get one that takes the correct number of shafts for your hemisphere. The unit on the left is a single-pin version while the one at right fits a four-pin hemisphere.

You'll pay around $100 for a mini spool or perhaps $140 with replacement pins. Once the axles have been locked together with a good mini spool, the next thing to break could be the cast-iron carrier or the pins themselves. Andrew said this is more likely to happen with a single-pin unit so he doesn't like to use them.

Although mini spools do a good job at a reasonable price, their limited size means only 28- and 31-spline axles will fit into them. If you want to go to 33-, 35- or 41-spline axles, the answer to your problems is shown at far left in **photograph 1** in the shape of the full spool. This replaces not only the pinion gears and side gears but the whole cast-iron carrier. Because there are no pinion shafts to break and because full spools are solid units, they are much stronger. Full spools are also fitted with bigger-than-standard bearings to take larger axles that are capable of absorbing extra power.

Different-sized bearings are available.

As Andrew points out, however, power alone is not the deciding factor. The way in which the car is set up is what really matters — 500 horsepower through a transbrake is much more severe than 500 horsepower through a foot brake.

In **photograph 2** the rear bearing cup fits the unit shown and measures 3.250". The bearing cups in front are factory units and came in the two sizes shown (3.062" & 2.891") with larger sizes used in heavier applications. The internal diameters of the centre races for the two factory bearings are the same at 1.781".

Photograph 3 shows a full spool fitted with its ring-gear. The aftermarket version has a larger internal diameter to take larger-diameter axles.

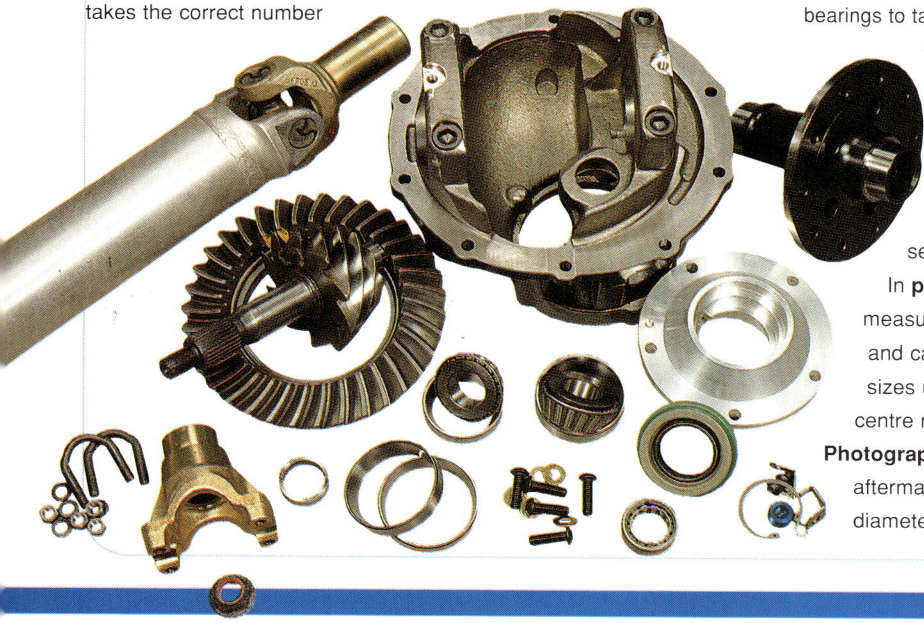

A full spool fitted with its ring gear.

THE RIGHT SUPPORT

Pinion supports are another weak area in higher-powered cars. **Photograph 4** shows three types. The two at right are different factory units with different-sized bearings. At top right is the heavy-duty type used in the GTHO, known as a Daytona type. Front and centre is a standard unit and at upper left is an aluminium aftermarket type from Moser designed to handle much greater power levels than standard units.

Three types of pinion carrier.

Photograph 5 shows difference in size between standard and high-performance pinion bearings.

The need for a larger size is easy to understand when you stop and think about what the pinion is doing under power. As the gear turns, its teeth are pressing against the teeth of the ring gear with enormous pressure. According to Newton's third law, the teeth on the ring gear are pushing back with equal pressure and trying to push the pinion out the front of the pinion support. It's these bearings that resist that force.

The difference between standard and high-performance pinion bearings.

As you'd expect, the front of the pinion supports are different on the standard and heavy-duty factory units. The Daytona is on the right in **photograph 6** and you can see there's more metal between the bolt holes and there's a better radius leading up to the seal area. The Moser is considerably different from both OE units. The size of the front bearings in all three supports is the same because rather than have a load-bearing function, its main job is to set the correct amount of pre-load, or resistance to turning, of the pinion shaft.

Front view of different pinion carriers.

A full spool, ring gear and pinion fitted with new bearings and ready to go.

Photograph 7 shows a full spool, ring gear and pinion fitted with new bearings and ready to go. Once we reach this stage it's time to talk about pinion yokes.

Photograph 8 shows some different types. There's a long and a short version and each of them can have a wide or narrow width. The width is measured between the universal joint locating tabs.

The two rear units are about an inch longer and the wider unit at right is the typical type used in heavy-duty applications. It takes 1330 type unis and is 92mm wide with $1\frac{1}{8}$-inch cups. This yoke is typically what you'll find in a mid-'70s Galaxie. Andrew says it's capable of taking 700 horsepower or more, depending on how the car is set up.

The other longer but narrower unit is typical of Australian nine-inch units. It takes a 1310 type uni and measures only 82mm across. At front right is a wide but short unit typical of an F100. Front left is an 82mm unit typical of a tank Fairlane and is popular for conversions because its $1\frac{1}{13}$-inch cup size makes it a perfect match for a Salisbury-type tailshaft. Although they can save a few dollars, Andrew doesn't really like using them because they're about the weakest yoke you can use in a nine-inch. Front and centre is a Moser yoke which is definitely the strongest shown.

Different types of yokes.

TOUGHENING UP THE FORD NINE-INCH

9

The assembled pinion gear and support.

HOW TIGHT IS RIGHT?

Photograph 9 shows the assembled pinion gear and support. There's a spacer between the front and rear bearings on the pinion shaft, which limits how close the bearings can be forced together when the nut securing the yoke is tightened. Looking back at **photographs 5 & 6** you'll see from the bearing cups in the front and rear of the supports that the bearings oppose each other. The factory method of setting how close they come to each other (and thus how much pre-load there will be on the shaft) is to fit a crushable washer that deforms a certain amount when the nut is tightened to the correct tension.

10

The full spool/ring gear assembly and the pinion assembly ready to have the engagement pattern checked.

The problem with a crushable spacer is that over time it gets squashed more than it should by the ring gear pushing the pinion shaft forward. The solution is to fit a solid spacer that will maintain its length. The correct pre-load is achieved initially by adjusting the length of the solid spacer on a linisher and then fitting it and checking the resistance to rotation. This procedure has to be repeated until the pre-load is correct.

Photograph 10 shows the full spool/ring gear assembly and the pinion assembly fitted into the centre and ready to have the engagement pattern between the teeth on the ring gear and pinion shaft checked. We described how this was done a few issues ago, so we won't repeat ourselves here.

The assembled pinion gear and support.

RIGHT IN THE CENTRE

Almost everyone will have heard there are different centre cases available. **Photograph 11** shows front views of the three main types. At left is a standard case with only one vertical rib. Not surprisingly, this is known as the single-rib version and is found on single-wheeler diffs in standard power applications. Next is the version known as the twin-rib version, for obvious reasons. At right is a Moser unit which, in addition to being twin-ribbed, is made from highly

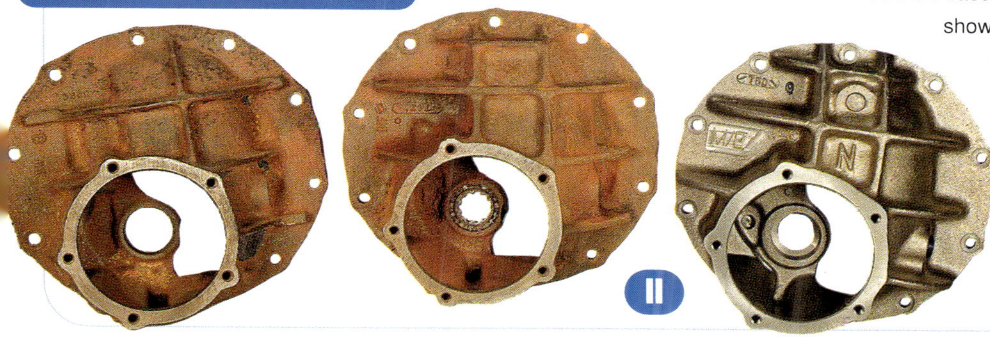

11

TOUGHENING UP THE FORD NINE-INCH

nodular iron. You can see the N, which indicates high nodularity between the twin ribs. This is the strongest iron type available. Ford also brought out a high nodularity case and it also has an N in the same place as the Moser shown.

The bearing retaining caps for the ring gear carrier also come in a scale of strengths. At left in **photograph 12** is a standard unit with cast-iron caps. There are also factory high-nodularity units available, which are marked with an N like the centres

12

The bearing retaining caps for the ring gear carrier also come in a scale of strengths.

discussed previously. In the centre is a standard iron centre fitted with much stronger billet caps. This is necessary because, as mentioned, the pinion gear is trying to drive the ring gear out through the rear of the housing. The only things that stop it doing so are these bearing caps. The Moser unit at right is fitted with forged caps which are the strongest types available.

Photograph 13 illustrates clearly the heavier construction of the aftermarket unit. Look at how much extra metal there is around the pinion pilot bearing mount.

This about rounds out our discussion of performance upgrades to a locked centre nine-inch diff centre. There are certainly other things that can be done but we'll talk about them another time. If you want more information about nine-inch diffs in general, give Andrew a ring on (03) 9793 1915.

13

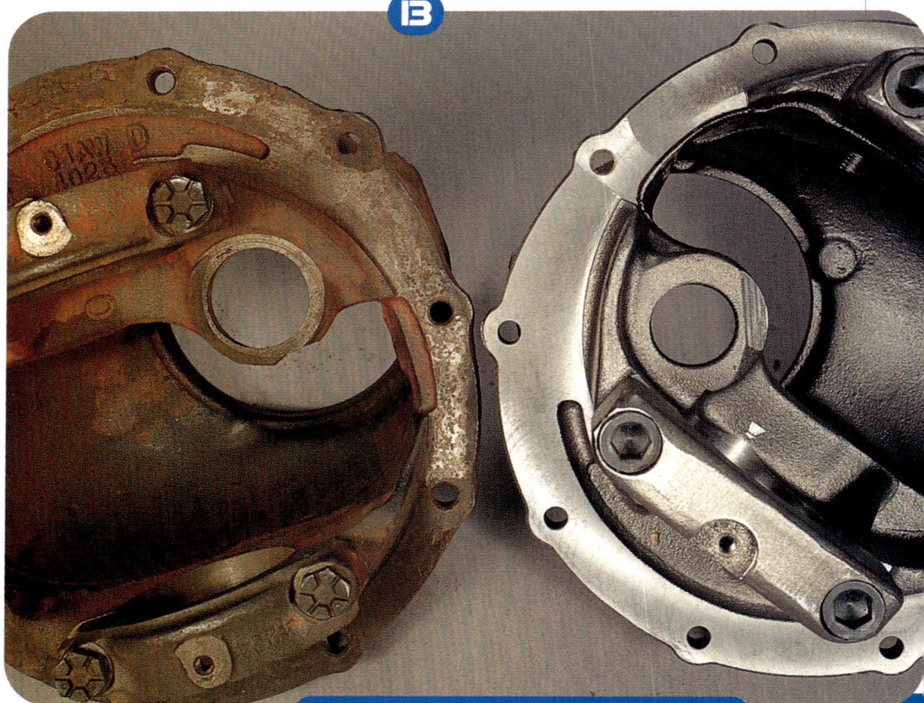

The heavier construction of aftermarket units.

INSERTING STRUTS

Front end feeling a little spongy? Maybe it's time to replace the strut inserts!

Fitting new shocks might be easy for some, but the strut is a different proposition. Rear shocks can, depending on the car, take about half an hour to change, but try fitting strut inserts without the right tools and you might just find yourself down the shit chute.

The following covers in detail what tools are needed and how to go about fitting strut inserts. The chosen vehicle, a Commodore, is one of the most difficult and therefore we have covered every possible obstacle.

Thanks to:
We would like to thank Brad at Traction Tyres and Suspension for his valued assistance. Brad can be contacted on (03) 764 2811.

1

2

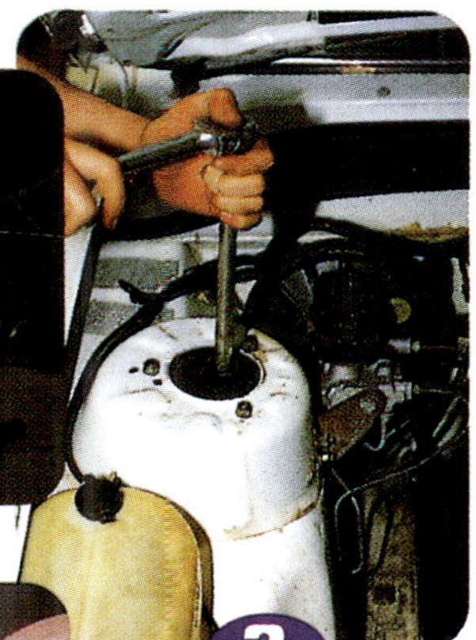

3

4

5

1. Start by opening the bonnet and marking the strut towers with a black pen. The mark should correspond with one bolt, the head of which needs also to be marked so that misalignment won't occur when refitting. This is critical.

2. Use screwdriver to remove plastic strut cap.

3. Using a 22mm socket, loosen (don't remove) strut top nut. Otherwise, once the strut is ready for removal, without the benefit of vehicle weight stopping the strut shaft from turning, it may be difficult to undo.

4. Jack up vehicle and place on quality stands. Good quality stands ►

Story/photos: PETE 'SPANNERMAN' DUNCAN

6

7

8

9

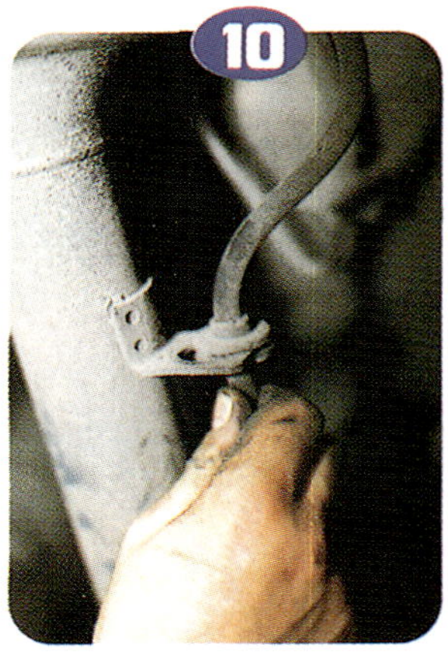

10

12

11

13

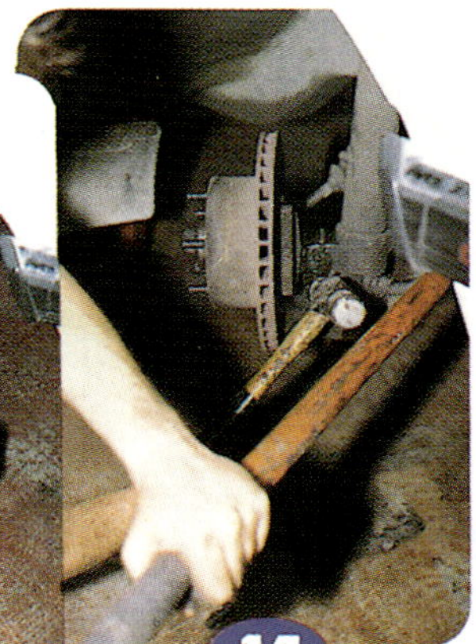

14

15

16

are available from most automotive outlets. Place the stands under chassis jacking points or on the chassis rails — not on the floor.

5. Once the wheel is removed, remove split pins from tierod ends and ball joints on both sides.

6. Using a 3/4" or 19mm socket, undo tierod end nuts to within last two or three threads.

7. Use two hammers to vibrate (shock) steering arm taper loose. Whatever you do, DON'T hit the tierod. Rather, place a hammer on one side of the rod end (stationary) and hit the other side. The nut is left on as a precautionary measure in case the hammer slips, hits and destroys the thread.

8. Once removed, swing the tierod arm away from the caliper.

9. Using a 19mm socket, undo and remove caliper bolts.

10. Before and during the process of caliper removal, twist brake line to a position which allows it to slide out of the strut body retainer. Using wire, support caliper out of the way.

11. Soak threaded section and taper on ball joints.

12. Use a 22mm socket to remove ball joint nut. The most difficult part of the exercise is breaking the ball joint taper. The easiest way of doing this is to use a Holden tool, however as these are not available to the general public another method must be employed.

13. Option A: ball joint. Use a forked tool which is rammed home using a large hammer. Although a fast method, you end up with a stuffed ball joint boot.

14. Option B: ball joint. Place a long bar between the crossmember and caster rod and lever down while slamming ball joint taper with round-headed hammer. Brad from Traction Tyres and Suspension advised that while this is a 'bushman's method', it's effective and means you don't have to buy new ball joint boots. Be careful what you hit, and DON'T hit the top of the ball joint!

15. Once you have broken tapers on both struts, use a long bar to lever down the caster rod so as to relocate it away from the strut.

16. Use a 13mm socket to remove three strut head retaining bolts while holding the strut leg to stop it falling on your foot!

17

18

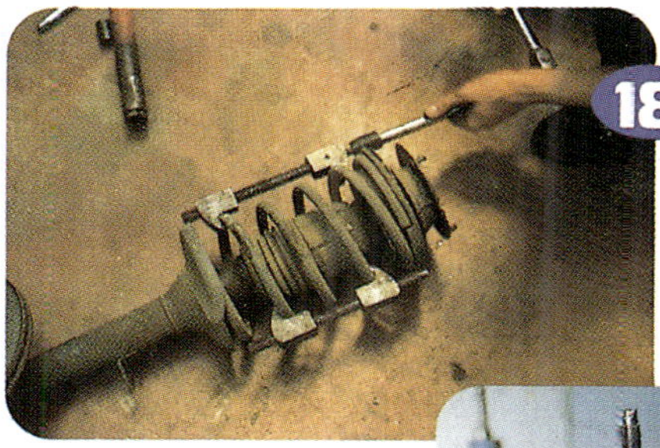

19

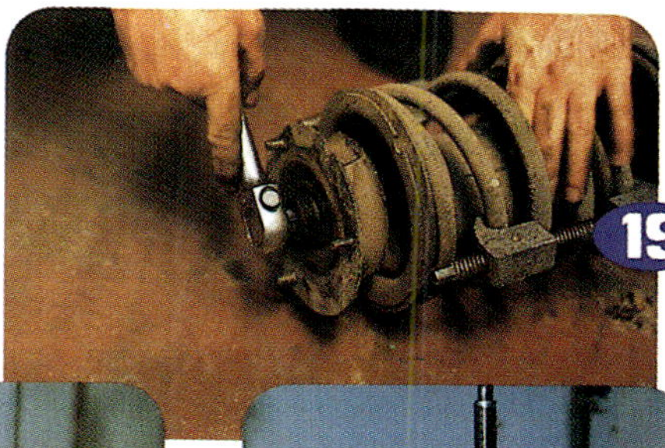

20

21

22

23

17. Strut leg ready to be dismantled.

18. Use a good-quality coil-spring compressor (available from most automotive outlets at prices ranging from $40 to $180) to hold spring while dismantling. Place adjacent to each other, using as wide a setting as equipment allows. Compress the spring until the strut inner is loose. Beware — the standard hydraulic struts may compress with the spring. EXTREME CAUTION must be used when fitting spring compressor; if poorly fitted the spring may displace like a rocket and take your head with it!

19. Use a 22mm socket to remove strut top nut (the last two or three threads) then remove spring (placing it carefully on the floor) top plate, rubber boot and bump stop.

20. Place bare strut leg in a vice and loosen large strut body top nut (gland nut) using a pipe wrench while trying not to mark the strut body.

21. If this is the first time the strut has been removed it will contain a 'wet leg' full of oil. Use a small screwdriver to dislodge and flick rubber seal upwards.

22. You can now remove the internals, which will slide straight out. Remember that the strut is filled with oil and unless you are careful it will go everywhere. Once the wet leg is removed, upturn strut and remove remaining oil.

23. Once the strut leg is cleaned (optional), simply slip the new insert in place and refit the gland nut. NOTE: gland nut must be very tight.

24

25

26

27

INFORMATION NOT RELATING TO PHOTO: Reverse the procedure, but when replacing the spring, make sure the spring end fits correctly in seat. Also, assuming top spring retainer is still in place, ensure that rubber dust boot containing the bump stop is fitted correctly, otherwise you may find the bump stop has dislodged and is slopping around in the dust boot.

24. Should spring retainer fall apart, there is a mark to show where it should be positioned — that is, towards the stub axle.

25. Fit top bearing plate and new strut top nut. DON'T hold the strut shaft with vice grips. Remove spring compressor.

26. Make sure you refer to the marked nut when placing strut in position, then reverse removal procedure and, once complete, fit wheels and place car back on the ground. This way you should be able to tighten top nut; if not, ask a local garage for the use of a rattle gun.

27. Aftermarket adjustable urethane bushes make caster and camber changes easy.

Hopefully, you have completed the job without any problems and, assuming you are fitting quality strut inserts, ride should be considerably improved.

GETTING IT DOWN

Putting your power where it belongs — on the road!

Story/photography: Ewan Kennedy

Nowhere does the word 'compromise' come into the car scene more than in the suspension department.

It's obvious that grip has to come into the equation, but you also have to look at ground clearance, car clearance (not rubbing wheel arches, etc), steering feel, drivability, tyre wear, comfort, not to mention the ever-present bogeyman — cost.

We called in to talk to our old mates at East Coast Suspension at Kirrawee, south of Sydney, (02) 545 1044, and spoke to head man Dave Falson. Dave's been involved in circuit racing for donkey's years — a racing Commodore and Porsche were in the shop while we were there — and he also does a lot of suspension mods on street cars, both current models and street machines.

Dave had an XB four-door coming in — an ideal subject for this feature. Its body was a little rough but had almost no rust, the V8 had already been modded and it ran a four-speed. In short, it was a typical tough Ford in the build.

The owner was on a tight budget and asked Dave to keep costs to a reasonable minimum, but at the same time he wanted decent handling and traction. Compromise, remember!

Do a complete check

"The first thing we do is look over the complete suspension," said Dave, "to work out what condition it's in and to see if anything had already been replaced.

"In a car of this age we might find that the shocks are relatively new. No car can get to 20 years of age without them having been replaced several times — anything over 60,000km on a set of shocks and they're starting to live on borrowed time.

"There's also a chance that the suspension bushes have been replaced on an older car. It's essential to check the condition of all the ball joints and make a note of the wheel-alignment settings. In this XB the steering box had been overhauled and the shocks weren't all that old and were still in reasonable condition.

"In a budget suspension job there's no point in spending money if it has already been spent. I recommend to your readers to find a suspension guy who takes that attitude and who is willing to sit and talk before work begins.

"It's smart to have a complete plan for the car and to do everything in one hit. Naturally, that costs a fair bit of money at the time, but it's often cheaper in the long run because you aren't throwing out some recently purchased components when you change your mind later."

Spending time in the bush

The XB's owner was out of luck — the car needed new bushes. To stiffen the car up, Dave recommends fitting polyurethane bushes at the front radius rods and the idler arm mountings. They can also be used at the lower inner control arm and the sway bars, but they're of less importance, so stock Ford parts can be used to keep costs down.

Which way to point?

At this stage Dave mentioned that, even if all the suspension bits were in good condition and didn't need replacing, he would suggest altering the wheel alignment. "Adding more caster and reducing wheel camber will increase the car's high-speed stability."

Shocking news

How much money do you have? A good set of Australian-made Monroe oil shocks will cost around $110 per pair. These bring the car back to standard. "Koni Special D shocks are good gear," said Dave. "They

Before you drive in, have a good idea just what you want from your car. A good suspension specialist will be willing to listen to you and make suggestions.

Checking the XB's wheel alignment settings. Going for different from stock front-end geometry can make quite a difference to the way a car handles and feels.

Reset leaf springs for the rear of an XB Falcon.

have been around for a long time and have proven themselves. A big advantage is that they are serviceable.

"Spending $220 for a rear pair and $300 for a front pair gets you the standard Koni Special Ds, which can be revalved to different settings for about another $80. You can use one setting on the road, say 70/30, and others — 90/10 is common in drag cars — at the track.

"It takes time to alter them, but if you can afford to go to $380 for the rear and $450 for the front you can get externally adjustable Koni shocks which can be reset in minutes. They're not cheap, but they really are good value if you want to get serious about racing."

Springs and bars

At the rear, Dave suggested resetting the spring leaves ($180) and probably adding an extra leaf ($100 more). The more you lower the rear springs the more important it becomes to use an extra leaf — that's because the lower springs have less curve in them. Don't use lowering blocks as the U-bolts can snap and, in any case, you get more axle tramp under hard acceleration because the springs are flatter.

"You could add a rear swaybar," Dave said. "I'd use an 18mm one, though it may not make any real improvement. You're better to fit a stock traction rod as used in the Falcon coupes and GTs of the period."

For street use the front end is more critical than the rear. Lowered springs — *not* cut springs — give improved road grip but reduce ground clearance and affect ride comfort. A 24mm front swaybar makes a significant difference (the standard bar is 20mm), but going for a larger size actually reduces road grip because the front is too stiff.

Another useful hint from Dave Falson: "Unless you are chasing the ultimate in grip it may not be worth while going for new swaybars. Simply fitting better springs, shocks and bushes may stiffen up the car enough to suit you. Try it and see."

RETREADED SLICKS COULD BE THE ANSWER

Flex Tred in Bankstown, west of Sydney, has been quietly retreading slicks for years and no one seems to know about it. Owner Jim Bridgeman reckons those who are using his tyres are running such good times they're keeping quiet about their tyres! He reasons that retreaded slicks heat up more quickly than new tyres so get better traction.

Simply take your tyre into Flex Tred at 28 Cottam Avenue, Bankstown. Or if you are from out of town ring them on (02) 709 5746. They'll tell you of a local tyre place to take it to be sent to Flex Tred, or you can ship it direct.

The old tyre can be either a slick or a street tyre, but it must be in good condition to be retreaded.

All but the very fastest drag cars can use retreaded tyres and Flex Tred also does work on speedway, rally and circuit-racing tyres.

The cost is $12 per inch of tread width.

Lowered Pedders springs. *Do not* cut existing coil springs.

before you make a move. Do you simply want a good street car with precise steering and good grip? And you drive it every day so you need some comfort.

Or do you want a full-on street machine with belly-to-the-bitumen clearance and big-name suspension components?

Or maybe you want to go drag racing on Sundays and use the car for commuting for the rest of the week?

Or enter in the Bathurst 12-Hour race?

Or is it all of the above? Tough luck. As we said at the start, compromise is the name of the game. And which way the compromises slant is something only you can decide.

Removing the stock front swaybar before fitting a new 24mm item. On an XB the front bar makes a lot more difference than does a rear one — you may be better off putting your money into a Ford traction rod rather than a rear swaybar.

Polyurethane bushes (these are Super Pro) stiffen up the suspension. They may not be necessary at all locations — see story.

Retreaded slicks are cheaper and many claim they give better traction as they heat up more quickly than new tyres. This one is a 275/60x15 — you'll get plenty of traction from a slick of that size!

steering to BIG savings

By Paul Tuzson

DOING a reco on a steering box is a lot simpler than most of us realise. Save money here and you can put the cash to good use elsewhere, like a bit more grunt from the engine...

IF YOU OWN AN OLD HOLDEN, sooner or later you're going to have to replace the internal bits in its steering box. They last well, but nothing lasts forever. The reco is a relatively simple job and there are real savings if you do it yourself. That way, your hard earned can go into more exciting automotive things. Like a gruntier motor, for example!

It goes without saying that there are no new steering parts for old Holdens, so it falls to aftermarket suppliers to fill the niche.

Internals for old steering boxes is just the sort of thing that Rare Spares specialises in, so we went and asked them how to go about the job.

Frank, their steering box guy, said that surprisingly few people know just how simple the job of overhauling the Holden box really is. He said that he'd even had fairly experienced amateur mechanics watch him do the job, and then kick themselves for having paid someone else to do it!

When reconditioning a Holden box the only parts that Frank keeps are the housing, cap and steering-tube. Everything else is stripped out and junked.

The remaining components are then hot tanked to get them looking immaculate and working to the max.

If you're going to do the job yourself you can either pay someone to hot tank them for you or you can clean them manually that's not a hugely messy or lengthy job.

Fire up the Barbie!

When all the bits are thoroughly clean you'll need to fire up the BBQ and cook them for a while! You see, aluminium expands more than steel when heated. Cooking the housings and end caps means that the ballraces for the bearings will simply fall out. There is one ballrace in the main housing and one in the end cap. (photograph 1).

Next, the old bushes in which the sector-shaft turns may have to be pressed out (photograph 2). This brings up the familiar old difficulty for the home mechanic where do you get a press? If you don't have a spare hydraulic jack lying around to make an inverted press, you'll just have to take the housing to someone and have it done for you.

The bushes and all of the other parts you'll need to replace are available from Rare Spares as a kit. Each item in the kit is also available individually (photograph3).

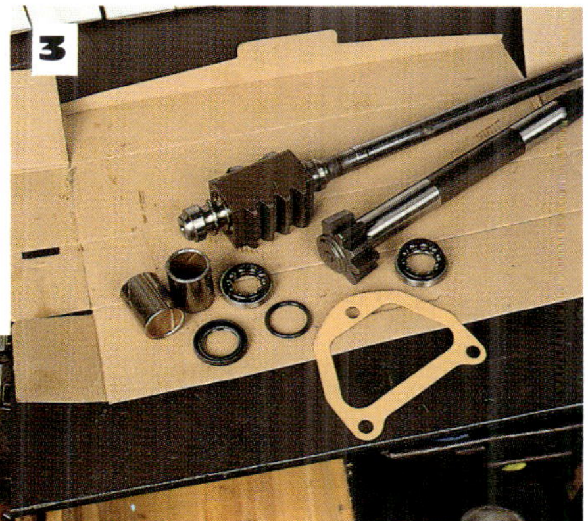
3

1

The replacement bushes need to be pressed in. This looks exactly the same as removing the old ones (photograph 2).
If new bushes are fitted they must be reamed to size (photograph 4).
An adjustable reamer is

4

used for this job but if you don't have a press you probably won't have an adjustable reamer, either. You'll have to get one from somewhere and set it to match the diameter of the sector-shaft. Perhaps this point will be enough to prevent you from doing the job yourself.

Fortunately, the sector-shaft bushes are often not worn as much as the other components. If this is the case with your box, you can probably get away with not doing the bushes at all which means that the hassles with the press and adjustable reamer won't concern you and the job can definitely be done in your garage.

Now, the bearing races need to be fitted. One is knocked into place in the end cap and the other is knocked into the housing in the same way (photograph 5).

2

5

The kit comes with a seal for the sector-shaft and this must be knocked into its seat in the main housing, taking care of course to make sure it seats squarely and to avoid damaging it.

One of the bearings is placed on the steering shaft which is then inserted from the end of the housing as shown (photograph 6).

The other bearing should be fitted into the end cap and retained in position with grease. If you don't make sure that it remains firmly located with grease during assembly it could fall out as you screw in the end cap with the result that you won't be able to turn the steering at all (Photograph 7).

Make sure that you hold the steering shaft horizontal with your fingers as you tighten the end cap until the shaft has no free-play but still turns freely. When it feels right, tighten the locking-ring.

Align the worm and block with the sector-shaft bushes by turning the steering-shaft. When the worm and block is centralised, support it with one finger so that it will engage properly with the sector-shaft (Photograph 8).

The original Holden boxes came with a choice of two ratios: 20:1 and 16:1. Rare Spares decided to split the difference and supply components which provide an 18:1 ratio.

Frank describes it as "runny grease".

The steering-box is designed so that there is greater resistance to turning when the steering wheel is centred. As the wheel is turned away from the central position, steering becomes easier. The final adjustment is for end pressure on the sector-shaft which controls this resistance to turning.

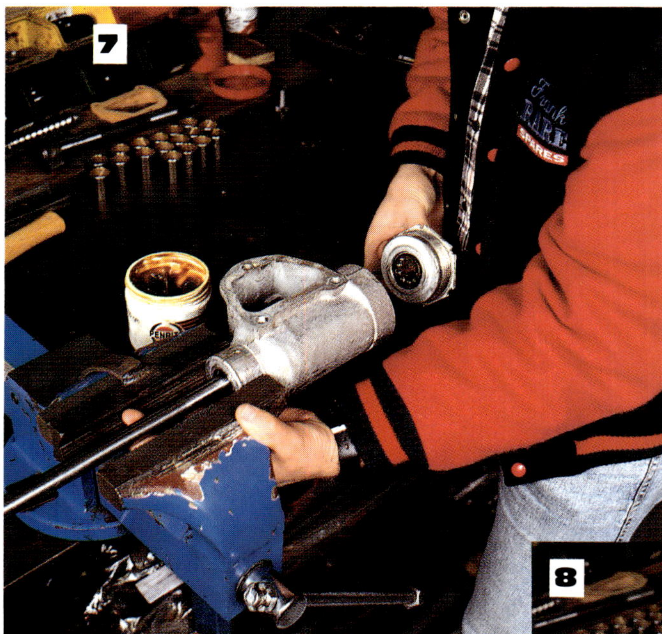

Oil Up

Fill the unit with oil, place the gasket on the main body and fasten the cover securely (photograph 9). Reconditioned units are shipped without oil so none is shown in the photograph. Any old oil will not do. There is special steering-box oil which is so thick that

When you centred the worm and block and inserted the sector-shaft, the steering-box itself was automatically centred. To remember where this position is, clamp a pair of multi-grips on to the protruding end of the sector-shaft (photograph 10). Having done so, you'll know the box is centred when the multi-grips are in the same position as when you started.

Because the teeth on the sector-shaft are bevelled, increasing the pressure on the end of the shaft causes them to interlock with those on the block more firmly. Tighten the adjuster screw in against the end of the shaft (photograph 11).

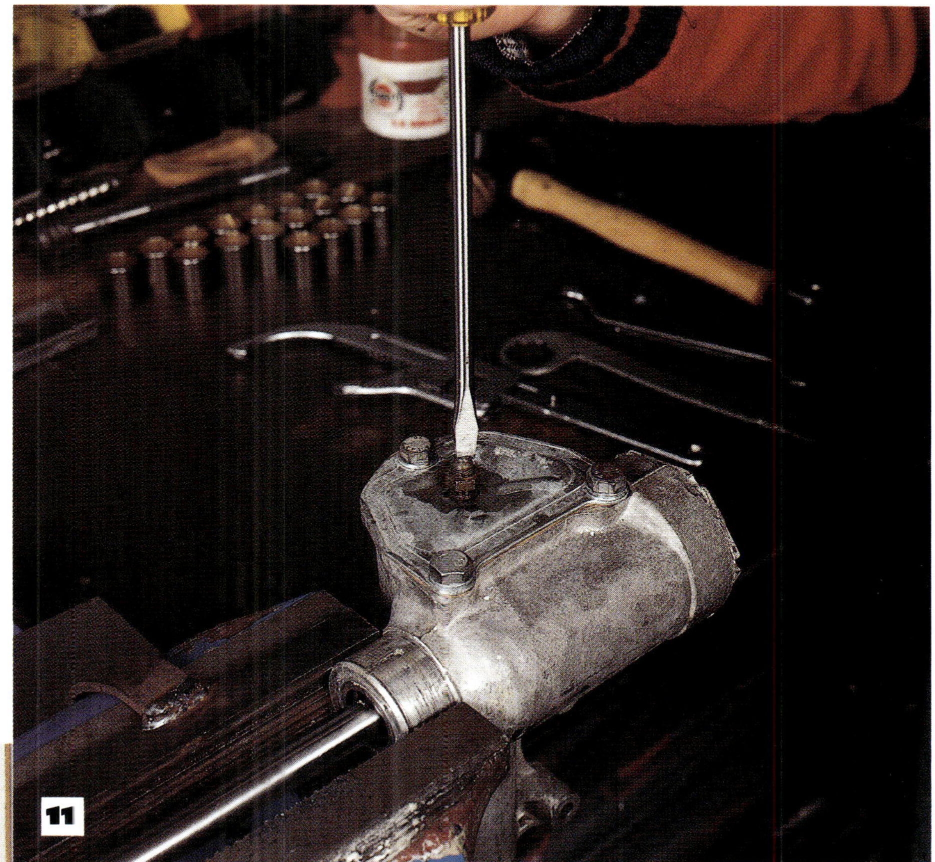
11

10

If you find that you can't turn the steering-shaft in manner shown in photograph 10 the adjuster is too tight. If the sector-shaft demonstrates any free-play the adjuster is too loose. The amount of pressure will be just right when you feel a firm but gentle resistance as the steering-shaft passes through centre. When it feels just right, tighten the lock-nut, re-fit the steering-tube, fit the assembly to your car and away you go.

As we mentioned previously, this box is available complete (less oil) from Rare Spares and they do plenty of boxes for other makes and models too. We'd like to thank Frank and the other guys for their time and knowledge which they freely imparted to us for this story.

All Barred Up!

Story/photos: JAMES GARROD

The Mighty Torana was endowed with better than average handling when it was released, but as they get older they can lose that edge and become downright dangerous!

The first thing average Torana owners do to their cars is fit a better set of wheels and tyres. Good move. This alone will improve road grip, handling and braking, but can, unfortunately, show up other weaknesses in the suspension and/or cause different handling problems.

As grip increases, the driver is naturally more confident and may corner that little bit quicker — then he notices an increase in the amount of body roll/lean. This lean is changing the weight balance of the car and can lead to a decrease in the size of the contact patch of the tyre (the bit of tread in contact with the road).

Energy and momentum is being used or lost when the body pitches from side to side, energy and momentum that should be pushing the car forward. Excessive tyre wear, especially on the outer edges, is another by-product of roll or lean. To control it, a swaybar is fitted, in effect a torsion bar which ties the suspension on both sides of the car together. As one spring is compressed during cornering the swaybar twists against the force or position of the suspension on the other side and body roll is reduced.

Most cars came equipped with factory front swaybars and some had factory rear bars. These factory bars, while adequate, were of a small diameter and mounted in soft rubber bushes which compressed drastically before any effect from the swaybars was felt. LC to LH Toranas had no swaybars at all, for some strange reason, but GMH fitted them to LX and UC Models. LC to LH models can be fitted with aftermarket swaybars very easily and LX/UC models can have their swaybars upgraded with a heavier duty item with polyurethane bushes instead of the less effective rubber ones.

Swaybar manufacturer Whiteline and suspension specialists Jim Hunter's

Suspension gave us the opportunity to witness the whole procedure on an LX Torana sedan. Full fitting instructions, all hardware and grease are supplied with the swaybar kits, but this step-by-step pictorial guide will show you how to do it before you try it yourself.

Front swaybar fitment

1

Loosen front wheelnuts, chock the rear wheels and jack the car up as high as possible. Support the car on jack stands (not bricks) and remove the front wheels.

2

Under the crossmember, to the left of the pic, you can see the factory (skinny) swaybar and sheet metal mountings. You will notice the 'factory' oil leaks that all Toranas have. These have a very detrimental effect on the standard rubber bushes making them soft and useless. Oil has no effect on polyurethane bushes so their service life is vastly increased.

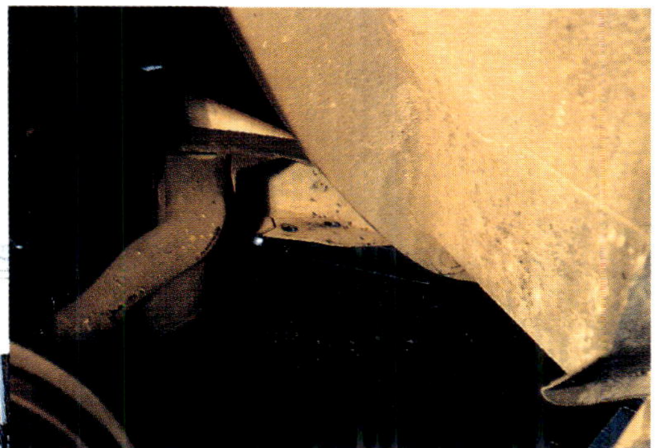

3

The factory bar mounts behind the wheels, below the engine, and the saddle mounts bolt to the two holes you see here. The factory rubber bushes have a moulded lug which fits inside these holes. The bolts supplied with the kit are smaller than the holes but a top reinforcing plate (supplied with kit) is of the correct hole size and sits above the subframe. LH Toranas do not have these holes in their subframes but full instructions and measurements to drill them are supplied in the kit.

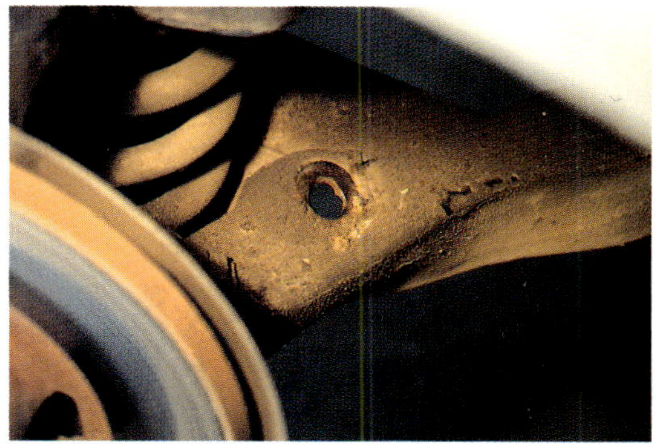

4

The mounting points for the bar's links are located at the rear of the lower front wishbone. Even LHs, which had no bars, have the holes. This pic shows wear to the link bolt-hole caused by sloppy, worn-out rubber bushes. The poly bushes supplied with the bar are flanged to fit inside this hole and as the wear was only negligible the wear was not repaired. If the hole is worn excessively then it can be arc or MIG welded and drilled/ground back to size.

5

Here the bar is assembled ready to fit. Note the heavy duty steel mounts, washers and lock-nuts, and the cad plating on all hardware. The bars are painted but can be repainted to suit your own tastes before fitting.

6

Before fitting the 'D' bushings (body mountings) to the bar, they must be liberally smeared with the black moly grease supplied with the kit. Slop it on real thick as you can wipe any excess off after fitment.

7

Trent (the wonder mechanic) lifts the bar into place and loosely bolts up the rear mounting on one side. He then loosely fits the mounting to the other side before attaching the links to the wishbones.

8

The link is then passed through the lower wishbone and the link-bolt nuts loosely nipped up. Note the flanged lower bush (mentioned in pic four) showing through the hole in the wishbone. The same procedure is repeated on the other side before tightening.

9

The bar is checked for clearance and alignment before everything is tightened. If any adjustments are required they are easier to make when the mountings are loose.

10

The link nuts can now be tightened down but *they should not be tightened so much that the poly bushings distort.*

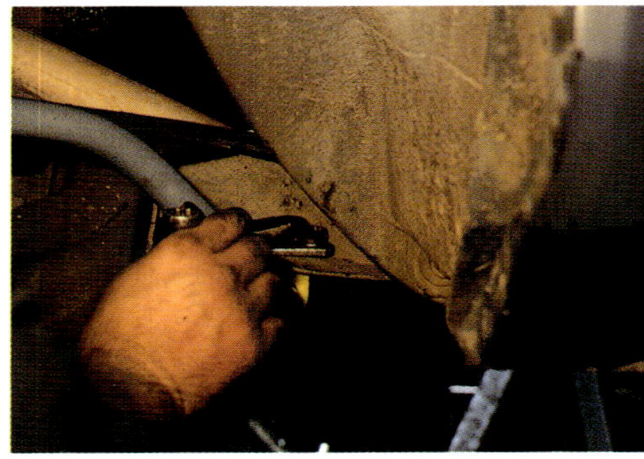

11

The rear frame mounts are now tightened down firmly. These mounts need a little extra torqueing but the poly bushes won't distort as the mounts crush up metal to metal. Don't forget to put the wheels back on before letting the car down off the jack stands.

Rear bar fitment

12

As with the front end, crack the wheelnuts, jack the car up as high as possible, support it on jack stands, chock the front wheels and remove the rear wheels.

13

The rear factory bars on LX/UC Toranas are as useful as tits on a bull. Notice the haste with which Trent is ripping it off. Don't be tempted to leave it in place (as some people do) as its limited effect is totally counteracted by the new bar. There is NO additional effect.

14

The Whiteline rear swaybar is partially assembled ready to install. The box sections on the ends of the links are the new mounts which have to be fitted to the chassis, since the bars use an entirely different method of operation (more efficient) than the factory setup.

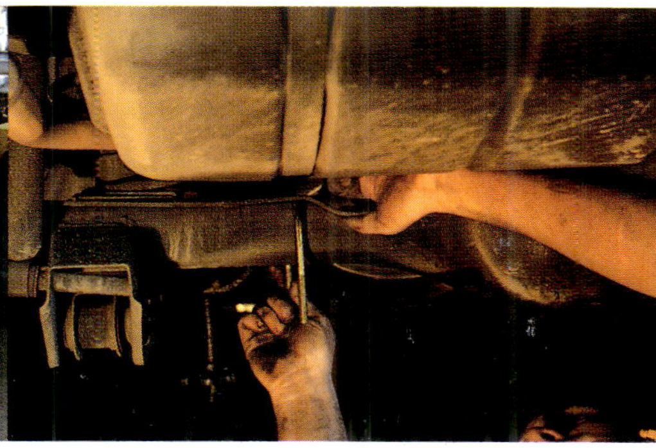

15

The new rear bar clamps over the diff housing with large U-bolts. These MUST be fitted under the brakeline to avoid crushing the pipe.

16

After lubricating the 'D' bushings with the supplied grease, the bar is loosely bolted up to the diff housing. The angle of mounting is most critical and the instructions have a very clear diagram showing the correct fitting angle.

17

Careful aligning of the mounting box to the upper, inner trailing arm mount is needed, then the bolt position is marked with a scriber or pencil. The bar is then swung out of harm's way and a centrepunch mark is made where the 3/8" hole is to be drilled.

18

This is why the car needs to be jacked up as high as possible — to get the drill under it. Always wear some sort of eye protection when working under a car, especially when drilling/grinding.

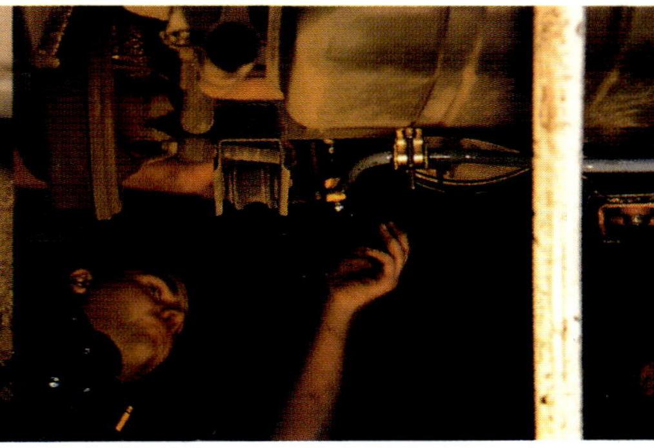

19

The instructions stipulate the position of the mounting box, which is bolted up very tightly to the chassis. The links are then loosely nipped up and the procedure repeated on the other side.

20

This extra view shows that the link is supplied slightly bent to align the swaybar to the mounting at the correct angle.

21

After checking all clearances and angles, and adjusting if necessary, the diff mounts can be tightened down firmly. The link-nuts are then tightened but, once again, not so much that the bushes distort. Put the wheels back on and let the car down gently. Do not forget to tighten ALL the wheelnuts.

The recommended retail price for Whiteline front and rear swaybar kits is $169 each. Jim Hunter's Suspension quoted $39 labour to fit each bar. Jim Gurieff, the suspension Guru from Whiteline, was most helpful with technical information for this article. For further queries or the name of your nearest supplier call (02) 820 2666.

Jim Hunter's Suspension of Auburn was also kind enough to answer questions and allow us to disrupt its workshop while we shot the pix. Trent Nowill (the spannerman) deserves thanks, and a medal for putting up with my jokes as he worked, and Ken Bowen helped out heaps with technical input on suspensions in general, and pricing. Jim Hunter's Suspension can supply this and many other products. Call 'em on (02) 647 1544.

From The Driver's Seat

I have owned my Torana for two years and still believe it is the best car I have ever driven. My only complaint came about as I was driving around the Mt Panorama racing circuit at Bathurst (on a club outing NOT the big race). The car handled like a boat on the twisty bits of the track, I nearly felt seasick. I replaced the stock wheels and tyres with 6" & 7" mags and Dunlop LeMans tyres, which improved the handling no end.

While I had much better grip from the new tyres, the car still cornered like a tugboat crossing Sydney Heads. The wallowing and lurching seemed even worse on some bends. Enquiries about solving the problem suggested upgraded swaybars would do the trick and the Whiteline brand name came up pretty often in my enquiries (the colour even matches my car). I decided to take up the challenge.

When I picked up the car I had no idea just how much difference the swaybars would make to my Torana's handling. I could feel the increased stablility of the car even reversing out of the car park. I pushed the car through gentle and sharp corners on my way back to work and even found that it negotiated roundabouts much tighter. The car now turns into corners as soon as I turn the wheel, instead of plowing on in an understeer situation until the tyres decide whether or not to grip. I can now take bends on or above the recommended speed limit.

The car feels entirely different, as I imagine it would have felt when it was brand new. I will now be upgrading the springs and shock absorbers as well as fitting a full set of polyurethane bushes to go along with the 308/Turbo-350 that is currently being built to go in the car.

Rebecca Patterson

More-than-happy Torana owner

GIMME A BRAKE

For serious street use, brakes are a lot more important than many drivers realise

Story/photography: JAMES GARROD

Leaking master cylinder rear seal, as evidenced by leak and rust marks on front of brake booster.

It's a sad fact that brakes are one of the most frequently neglected items on a car. They don't win you too many points in shows and they don't help your car go any quicker in a straight line. What good brakes can do is stop you dying ahead of time . . .

Common brake problems

The easiest way to avoid brake problems is to regularly inspect and maintain them. When you're washing your car have a look at the inside of the wheels to see if there is any fluid leakage. When checking your oil and water, check the brake fluid level and take note of any rapid fluid usage — this usually means a problem.

Make sure the car pulls up straight at all times, a flat road is best for testing as road camber can upset directional stability somewhat.

Does your pedal go low before any appreciable braking happens or do you need to pump the brakes for them to work?

Do you experience any grinding noises or pedal pulsations under braking?

How many times can you get power assistance from your power brakes (after the engine has been turned off) before the pedal goes really hard?

Does your hand brake work well, that is, can it pull your car up reasonably well from around 50 km/h?

If you are experiencing any of the problems mentioned stop now! You really shouldn't be driving your car until a thorough inspection of the brakes has been made by a competent person and adjustment or repairs have been performed.

Fluid Loss

Brake systems use hydraulic fluid to provide the clamping force to operate the disc calipers and/or drum shoes. The fluid is formulated to operate at high temperatures without boiling. The fluid should be clean and at the correct level at all times. The filler cap should be kept clean so that no foreign bodies enter the fluid; and it should be tight, with the sealing gasket in good condition.

If the cap is loose or is left off for any length of time the fluid can become contaminated by water from the atmosphere; it, hygroscopics, meaning that it will attract water and blend with it. If this occurs the internal surfaces of the wheel cylinders can rust, causing seal damage. Worse than this, the water in the fluid will lower the boiling point of the fluid and can cause brake failure.

The fluid should be completely drained from the system and replaced yearly so that it stays in good condition.

A small drop in fluid level is inevitable. As the disc pads and linings wear, the operating pistons in cylinders and calipers move in closer to the disc/drum. This causes the fluid level to drop and it must be topped up or air will enter the system.

If air is introduced, a soft pedal, one that needs pumping up to operate, or even total brake failure, can be experienced.

The first check for fluid leakage is a visual inspection of the brake backing plate inside the wheel. If an oil leak or oil stain is detected rectify the problem immediately or the linings can be permanently damaged.

Note that a similar leak pattern can be produced by a leaking axle seal The way to differentiate is to smell it. Diff oil smells like oil but brake fluid smells like squashed ants!

Even if no leak is apparent on the outside, a drums-off inspection is still advisable. Support the car on stands — not by bricks or the jack — chock the other wheels to prevent the car moving as the hand brake operates on the rear drums and they're bloody hard to remove with the handbrake on. Then remove the wheel.

Most drums will pull straight off, but if they have been adjusted up and a small lip has been worn into the drum surface then you may need to back the adjusters off to remove the drum. With the drum removed you will be confronted with a dirty, dusty interior which may show evidence of fluid leakage.

Locate the wheel cylinder, it's usually at the top of the backing plate, and carefully peel back one or both protective rubber boots. If there is any leaking or weepage, replacing or rebuilding is needed.

Not many people rebuild wheel cylinders as the cost of new replacements is usually quite low. The decision is yours.

The job will be easier or cleaner if you first clean the entire brake mechanism. This can be done with compressed air but it's not advisable. Some brake lining materials still contain some asbestos and even the smallest amount in your lungs can cause cancer.

The best method of cleaning is with water and detergent and a brush. Don't use degreasers or strong cleaners/detergents as they can cause damage to the linings. Dry everything off completely before starting work. This can be done with compressed air but be sure that there is no brake dust being blown about.

The next step is to take careful note of where everything goes *before* you rip it out. A workshop manual is recommended as it usually contains photographs or drawings of the completed brake assembly.

Carefully remove the retaining springs which hold the linings in place, taking care not to let them fly into the grass or into your eyes. Remove the small lining retaining clips and carefully remove the handbrake cables and operating levers. This may not be necessary as some linings will swing out of the way and hang on the cable without causing any damage.

Be careful if your brakes are fitted with self adjusters. These systems are quite complex and hard to reassemble. Some people remove these self adjusters completely. If

Wheel cylinder pulled apart for inspection or rebuilding.

Wheel cylinder being honed.

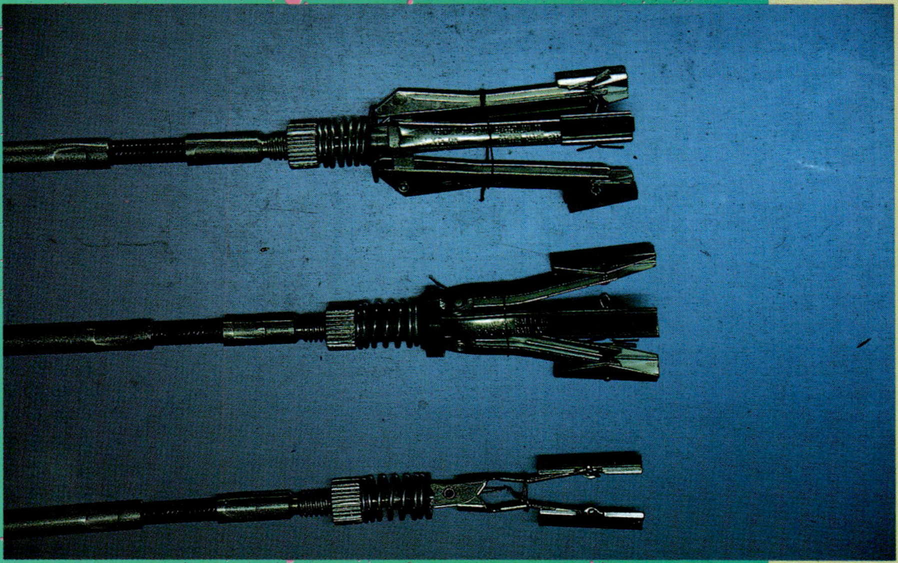

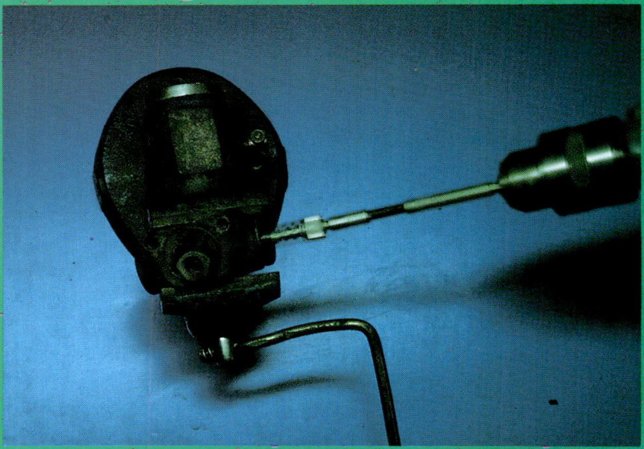

you do that it's imperative that you regularly adjust the brakes manually.

With everything removed you should now have just the wheel cylinders bolted onto the backing plates, note the plural — you should never perform brake work on just one wheel, always do both sides at the same time.

Using the correct size spanner, and there are special ring spanners made with a slot cut out to go over brake lines, undo the hydraulic line from the back of the cylinder and remove the retaining bolts that hold the cylinder in place.

If you are replacing the old cylinders with new ones you can now reassemble everything in the reverse order that you removed them. This is now a good time to inspect your axle bearings and seals. It only means undoing another four bolts.

You might like to pull off the brake backing plate and give it a coat of paint. After everything has been cleaned you can do a neat job with a spray can. If you're into showing your car this could pick up a few extra points.

Cylinder rebuilding

First remove the end boots, no need to be too careful here as you will get new boots in the overhaul kit, small steel or alloy pistons should fall out. If they don't come out freely it could mean they are rusted or corroded in place. Don't try to prise them out with a screwdriver as it is possible to score the cylinder's inner surface. The rubber inner-cup seals usually have a light spring holding them out against the pistons, push against one cup and the other will be forced out by the spring.

Clean and check the spring, any other internal metal washers or spacers and the cylinder body and pistons. They can be cleaned with degreaser or special brake cleaners but it's just as easy to use warm, soapy water. After a thorough drying, compressed air is perfect, check the bits; if any show signs of corrosion, rust or have any deep scoring then it is best to renew them. The pistons can be cleaned up with fine wet and dry in 800 or 1200 grit if they need it.

Mount the cylinder in a vice, not

too tightly, and get your electric drill and cylinder hone. Haven't got a hone? Your friendly accessory shop should have one for around 15 bucks. Some hones are screw adjustable whilst others use spring and/or centrifugal force to provide the surface pressure. The hone needs some form of lubrication to help float the metal particles away and stop the stones from clogging. You can use brake fluid but soapy water works well, a squirt bottle will help put the fluid where it does the most good.

Insert the hone into the cylinder with minimal contact on the surfaces. A variable speed drill is best but normal speed will do the job, start the hone spinning and vary the speed whilst moving the hone in and out for the length of the cylinder. Try not to pull the hone right out of the cylinder as it can grind the cylinder bigger at that point and/or damage the hone.

Keep the water (lubricant) flowing through and check from time to time to see if the surface is honed properly. It only needs to have the score marks and glazing removed, more than that and the cylinder may go oversize. A thorough drying out and you're ready for reassembly.

That process demands cleanliness so go inside and wash your hands. All the new bits go back in with a liberal coating of new brake fluid and it's not a bad idea to put a dab of rubber grease in the ends of the cylinders after the pistons are installed. The rubber grease lubricates the piston and helps to keep dust and water out. The new boots go on now and the brakes are then refitted and adjusted.

Sometimes it can be hard to find replacement cylinders, especially for older cars. But even a badly rusted or corroded cylinder can be resleeved by boring out and fitting with a stainless steel sleeve. The fluid entry/bleed holes are rebored then the cylinder is honed to the correct size. It's a job for the experts so don't try it yourself.

You'll be looking at about 50-60 bucks a pop. Whilst this may sound expensive, if you can't get anything else what can you do? The stainless cylinder liner will last longer and have a much higher level of corrosion resistance than the cast iron original.

Brake shoes and linings

Replacement of these has pretty well been covered in the previous section but there are things to look out for and a list of don'ts to observe. First check the brake drums; if they are scored or have a lip worn in then they will need machining. The manufacturers set a maximum diameter that a drum can be machined out to, sometimes the specs are cast into the drum so there can be no confusion. No specialist is allowed to machine past the max size or they will get in the poo.

The size is set so that the drum will still handle the load and temperature of normal use without cracking, warping or flexing. If you need a new drum, buy one — it just ain't worth the risk of bodging up a brake job.

Fitting linings or shoes is pretty straight forward but if your drums have been machined you must buy oversize linings to suit. The place that machines your drums should also match-grind your linings to the exact size. Once this has been done I like to give a slight chamfer to all sharp edges of the lining surface with a file or belt linisher (use a face mask with the linisher — remember the big C). This makes the drum go on a bit easier and helps the linings bed in a little quicker with a minimum of chatter and squeal.

Standard friction materials are good enough for almost every application these days. The only times you may need a heavier duty lining are for racing, carrying heavy loads or trailer towing. If this is the case you would probably be better off with discs on the rear; most cars can be converted relatively easily.

Reassembly of the brake system should be straight forward if you took note of the earlier comments. Bleeding air from the system and careful adjustment should have your drum brakes operating as well as new. Bleeding the system will be covered later. Adjustment will be needed within a few hundred kilometres as the new linings wear in.

Disc brakes

A similar inspection for fluid leaks as with drum brakes should be done first but discs operates on a different principle with higher pressures and temperatures. Place the car on stands, remove the wheels and check how much life is left in your disc pads. Anything less then three millimetres of pad means replacement is necessary.

In some vehicles, early Toranas for instance, the disc caliper is a dual-piston design. This has a piston and cylinder on each side of the disc which force the pads onto the disc surface. Most disc calipers these days use one cylinder on one side and a sliding caliper arrangement to clamp the other pad. This system is effective on normal cars but isn't ideal for high performance applications.

Scored and grooved brake drum.

The early system is also easier for pad replacement with two pins retaining the pads in the caliper and no need to remove the caliper to replace pads. Race pit crews can replace brake pads quicker than you can change a tyre. You should be able to effect a pad change in an early Torana in less than half an hour with only the most basic tool kit.

With the brake pads removed any fluid leaks from the cylinders will be obvious. The rubber dust sealing boots between the caliper body and the pistons are more delicate than their drum brake counterparts. They should not be poked or prodded with sharp metal tools. If they are still soft you can carefully place a small, thin metal object such as a smooth screwdriver between the piston and the seal to check for leaks. If any leaks are detected then it is advisable to buy a new or reconditioned caliper rather than attempt a repair yourself.

If you are competent (or just plain stubborn) you can remove the caliper, after disconnecting all lines, and firmly place it in a vice — use jaw protectors. Disassembly and compression of the pistons should be performed to dispell any fluid remaining.

The two cylinder calipers are bolted together with O-ring seals on the internal fluid passages. Rebuild kits will contain new O-rings and you should not contemplate reusing old ones.

The piston dust seals are now removed to expose the pistons. If the piston does not pull out easily, and a scored or marked surface inside the cylinder may make it hard, then special piston removal tools may be required. If you want to

Scored and grooved brake disc.

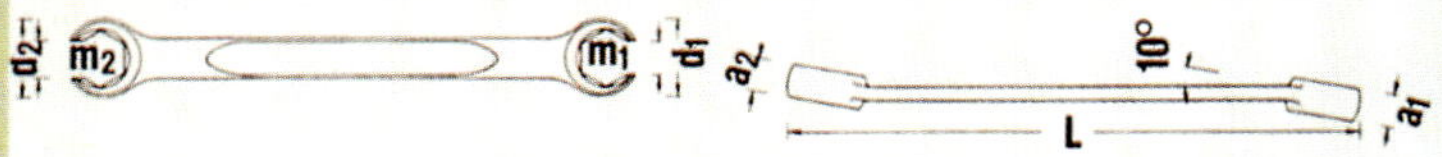

240a Double Ended Open Ring Spanner, Hexagon

AS-drive

AS-drive, Chrome Alloy Steel, chrome plated,
Fed. Spec. GGG-W-636E

Code	"	L mm	d_1 mm	d_2 mm	a_1 mm	a_2 mm	m_1 mm	m_2 mm	g
41621620	**1/4x5/16**	105	16.5	18	5.5	7	3.7	5	32

Application: on pressure pipe flare nuts

The right tools for the job. Slotted spanners for lines and nipples.

call it quits at this stage you can still exchange a partially stripped caliper; provided you give the brake shop all the bits you've removed.

Honing the cylinder may require a larger hone than that used for the rear cylinders but the procedure is much the same as already described. Some extra care must be taken so as not to bottom the hone out on the end of the cylinder. The pistons can be cleaned up with fine wet and dry, if needed, before the new rubber seals are fitted. Careful washing and reassembly of the caliper can now take place, followed by refitting to the stub axle/disc assembly.

Whilst disassembled the caliper could be polished (if alloy) or smooth ground and painted with a high temperature resistant paint to improve its looks. Chrome plating of brake or load-bearing suspension parts is not recommended due to the possibility of breakage from hydrogen embrittlement.

If the discs shudder under braking, are grooved or are glazed badly they should be machined. Once again we find minimum thicknesses set by the authorities. If your discs are too thin they cannot work properly, if they require too much machining to flatten out or true up you must replace them. Whilst the caliper is off the stub axle it is a good time to check, clean and repack or replace wheel bearings and seals.

Reconditioning single piston calipers is easier than doing a dual cylinder caliper but there is an extra area that needs attention. The sliding section of the caliper may just be a metal to metal V-shaped section with or without a metal shim between them. It could also have a hardened pin arrangement covered by a flexible boot.

With the caliper removed, or at the least the pads taken out, check for smooth operation of the caliper slide. If any stiffness is experienced then disassembly, repair, cleaning, or slide-rods or shims are required.

If rubber boots are fitted to the slides they must be replaced. Free sliding operation of the caliper slide should be checked on reassembly.

Before fitting new pads I recommend carefully cutting a groove, about 5mm deep, across the pad surface with a disc grinder (remember to wear a face mask). A hacksaw may not cut well enough and the groove it makes is not wide enough to provide any benefit. Some pads already have the groove in them, it's there to catch the brake dust and reduce brake squeal.

I also like to put a slight bevel on the edges with a file to remove any sharpness and allow easier bedding-in. At all times you should beware of sharp metallic fibres that are part of the brake pad material, they are worse than wood splinters.

Some disc calipers have thin steel shims fitted between the pistons and the pads. These anti-squeal shims put the pads at a slight angle to help reduce shudder and squeal, if they are fitted be sure to refit them. They have a small arrow cut into them showing which direction they should be pointing.

Refit the flexible brake hoses and any retaining clips, clamps or springs and bleed the calipers.

Boosters

In Toranas the power brake booster is fitted against the firewall before the master cylinder. It uses vacuum from the engine to pull a large internal diaphram forward thereby multiplying the pedal pressure.

There is a plastic one-way valve where the vacuum hose connects to the booster housing. This valve must be clean, undamaged and fully operational. Visual inspection is usually sufficient but if in doubt check with a specialist or replace it anyway — they are pretty cheap.

The vacuum hose should be clean inside and out and sealing correctly at all connections. Any cracking or perishing of this hose means the hose must be replaced. Only hose designed for a vacuum application should be used, certainly not heater hose or the like.

Booster problems are usually evidenced by the need for greater pedal effort needed or a slowly sinking dropping pedal when your foot is left on it. A quick check for booster vacuum retention is to start your engine then turn it off without touching the brake pedal. Then carefully pull the plastic vacuum hose valve out of the booster. This should take some effort but is not impossible and a hiss should be heard as air enters.

Worn brake linings due to lack of maintenance and/or adjustment.

If only a short hiss is heard, repeat the procedure (after restarting and stopping the engine) but wait 10 minutes before pulling the valve/hose out. If less or no hiss is heard there is a vacuum loss in the booster assembly.

Another test is to find a deserted road and turn off your ignition with the car coasting in neutral. If your car is fitted with a column lock don't engage this or you lock the steering. You should be able to stop the car or, at the very least, have decent pedal assistance for two or three applications before the pedal goes hard. If you don't, the booster needs checking and possibly rebuilding.

This unit is far too complex to rebuild at home, often needing specialised tools and jigs to pull apart and reassemble.

A brake specialist can sell you a rebuilt unit, they vary in price quite markedly with some models selling for $90 and others several hundred. Using second hand boosters will only save you money in the short term as it could possibly run out of guarantee and then fail a week later. Spend the bucks and get the job done right in the first instance.

The brake booster can be painted, polished or chromed but this should only be done when it is totally disassembled. The brake specialist should pull it apart for you, give you the bits to detail or chrome, and then put it all together for you. Whilst the booster is off you can clean or repaint your firewall in that otherwise hard-to-get-at area. How about detailing the booster mounting brackets as well?

Master cylinder

Most cars made since the 1970s have dual circuit braking systems, you can tell by the master cylinder having two caps and/or resevoirs for brake fluid. If you lose fluid or pressure to one system you should still have sufficient brakes to pull the vehicle to a halt.

In the case of non-boosted brakes if one circuit fails you can easily feel the poor performance of the brakes. It isn't so easy to detect a single-circuit failure with boosted brakes and we must revert to visual checks. It should be immediately evident if the rear section of your master

Double-cylinder disc brake caliper.

Sliding disc brake caliper.

cylinder is leaking by either excessive fluid loss or a leak at the rear of the master cylinder where it bolts to the booster.

If there is no external leak there may be one into the booster itself. The quick way of checking is to remove the vacuum hose connection from the booster and drop one end of a piece of clean string down the hole. If it comes out wet and smelling like squashed ants then the rear seal of the master cylinder is leaking into the booster. If fluid has got into the booster you will need to rebuild it.

Experience has taught me to regularly unbolt the master cylinder and pull the cylinder far enough away from the booster to make a visual check of the rear seal area. There is no need to disconnect the hydraulic hose fittings, so no air can enter the system.

Once again, check the cost of a new cylinder as it may be more economical to replace than rebuild. The entire piston assembly has to come out the rear and is held in place by an internal circlip. Some require circlip pliers to remove while others can be removed with flat pliers. The piston assembly should be easily removable but you may need to use needle-nose pliers. If it still refuses to budge then compressed air forced into one of the fluid exit holes may be needed. Be careful that the piston assembly doeasn't fly across the room and get damaged.

Thorough cleaning of the piston assembly is required before reassembly and inspection and cleaning of the cylinder's internal bore should be performed. Light honing of the cylinder is recommended, following the same guidelines as recommended for disc caliper cylinders.

Resleeving of master cylinders, as described in the wheel cylinder section, is also available but is seldom needed due to ready availability of new or rebuilt units. Careful cleaning of the master cylinder resevoirs is a must before reconnection; if any cracks, warps or splits are found the resevoir should be replaced. Under no circumstances should you consider reusing old brake fluid even for initial priming and bleeding.

Brake lines

Inspect the brake lines and fittings for leaks, cracks, dents or other damage. All flexible lines need to be supple and damage free, as should all joins and fittings. If any damage is found replace the offending item. Most big brake suppliers should be able to get any lines and fittings you need, if not they should be able to make them.

If any pressure switch or balancing or pressure-dropping devices are fitted they should be inspected and tested for correct operation. When you are satisfied that your braking components are all rebuilt, replaced or in good condition, then you can refit them to your car and bleed and adjust them.

Bleeding the hydraulics

If there is any air in your braking system, it either won't work at all or will not work properly. Before you attempt to bleed the brakes you need to keep a few things in mind.

Brake fluid will work like paint stripper if it gets on the car's paint. Place clean rags under the master cylinder. Brake fluid stings like hell if it gets in open cuts or squirts in your eyes. Safety glasses and gloves make a lot of sense, as do Band Aids on scratches.

Bleed nipples are small and can seize up due to exposure to water, road grit and mud. You must use the right size spanner on them. If they refuse to budge you may need to spray them with WD40 or a similar product. The aforementioned slotted ring spanners are also made in small sizes to fit bleed nipples.

Essential items when bleeding brakes are a large jar and a length of clean, clear plastic hose that fits tightly over the bleed nipples.

Start by topping up the master cylinder resevoir and putting the caps back on. A few pumps of the pedal will start the fluid flowing into the master cylinder. Now you will need a mate to help. Starting with the wheel furthest from the driver get your assistant to pump the brake pedal a couple of times then hold it down.

With the hose connected to the bleed nipple and its other end in the jar, release the bleed nipple while the assistant holds the pedal down. A spitting sound should result each time you open the nipple. Close the nipple quickly, to prevent air from getting back into the system. Under no circumstances should the assistant release the brake pedal

Piston-compressing tool for slide calipers.

How To FIT A LINE LOCKER

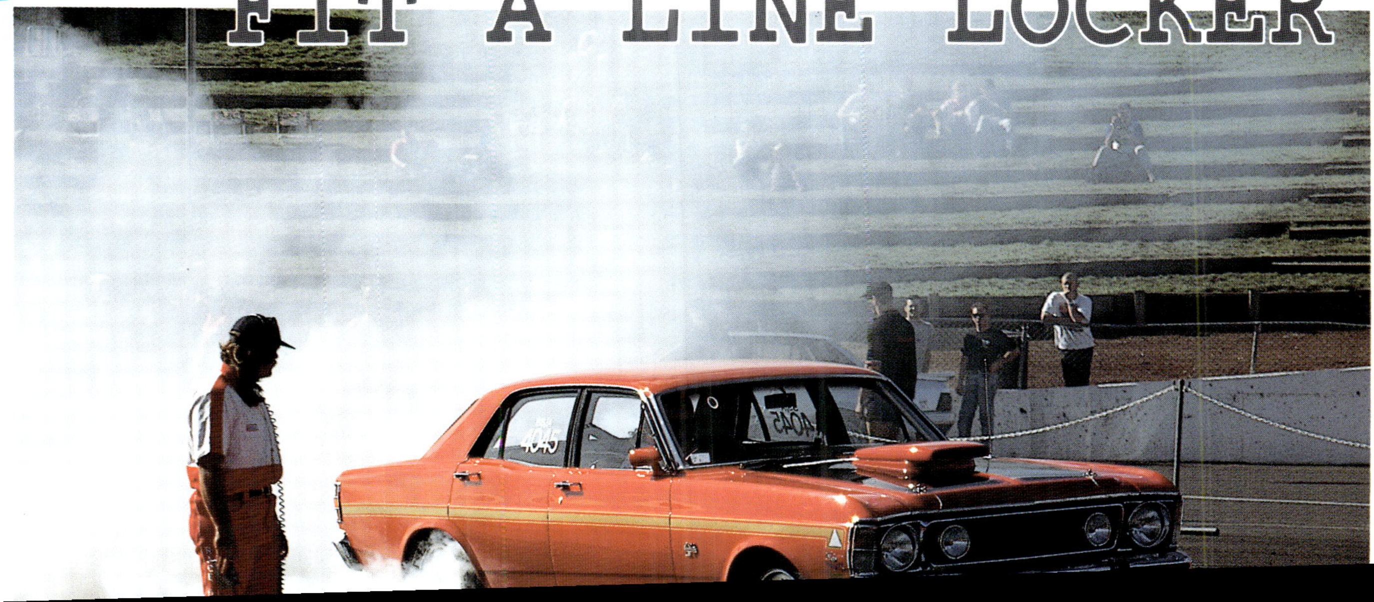

These brake linings have been ruined by fluid leakage — fit new ones.

until you have tightened the bleed nipple.

Repeat this process until they can feel the pedal getting harder and further from the floor.

If no progress is felt, you may not have completely drained your system of air, so you should move to the driver's side rear wheel and repeat this process. You should now be getting some feel in the brake pedal, albeit nowhere near as good as it is with all brakes adjusted and bled correctly. If not go back to the first wheel and start again.

In the case of power-assisted brakes the pedal effort will be harder than normal as you will have expended all the vacuum assistance. Keep this in mind as the brakes will not feel right until you have started the engine to provide assistance. Your assistant will need to keep a constant check on the master cylinder fluid level as fluid will be used up quite rapidly during this initial bleeding process. If it is allowed to run dry you will introduce more air into the system and the whole process will need to be repeated.

Once you are satisfied that there are no air bubbles, and only clear fluid is coming from the bleed nipple, proceed to the passenger side, then driver's side front brakes and bleed them.

Take care not to get any fluid on the discs or pads.

Adjustment of the rear wheel brake linings may be necessary to achieve the best pedal feel. This process varies from model to model so I recommend checking the workshop manual. There is usually a slot in the backing plate, sometimes filled by a rubber bung, that gives access to a toothed wheel on the adjuster.

By levering up or down against these teeth you will be screwing the brake shoes outward into the brake drum. A screwdriver can sometimes be used for this job but a special, bent adjusting tool can be picked up for a couple of bucks. Once the rear brakes have been adjusted you will also need to adjust your handbrake cable for correct operation.

Disc brakes do not need adjusting as they self adjust by retaining the piston in the correct position for minimum movement before brake activation.

Start up the engine to build up the booster vacuum; your pedal should feel as good as new or better than you've ever felt it. With pressure held down on the pedal, a quick visual check of all joints, connections and fittings should show any leaks.

Replace all the wheels remembering to tighten all the nuts. Top up fluid level and seal the cap down tight. A road test is needed to ensure the car pulls up straight and true but be careful. Even though the pedal may feel good and hard the brakes will not work at 100 per cent efficiency until bedded in. They may pull slightly for the first few kilometres but this should settle down. If they pull drastically to one side then you may need to adjust the wheel on the opposite side to the lockup or recheck the hydraulic circuit for air bubbles. If excessive rear wheel lockup is found then the front brakes may need to be re-bled.

Adjustment of drum brakes will be required within a couple of hundred kilometres to make up for the brakes bedding in. Checking of fluid level at this time is recommended as it is at at regular intervals in the future.

Fluid types

For normal everyday use and even the occasional excursion to the race track factory recommended brake fluids are fine. As mentioned earlier, all fluid should be drained and replaced yearly.

It's not recommended to mix different brands of brake fluid. Always buy more fluid than you need and store it, tightly sealed, upright in a dry area. There are high temperature, high silicon and racing type brake fluids on the market but they are expensive and not really necessary on a street car. They are not compatible with any other form of brake fluid so a complete rebuild and flushing of the entire braking system is needed before use. Some even require special brake seals as they do not work with conventional seal materials.

1

2

3

4

5

6

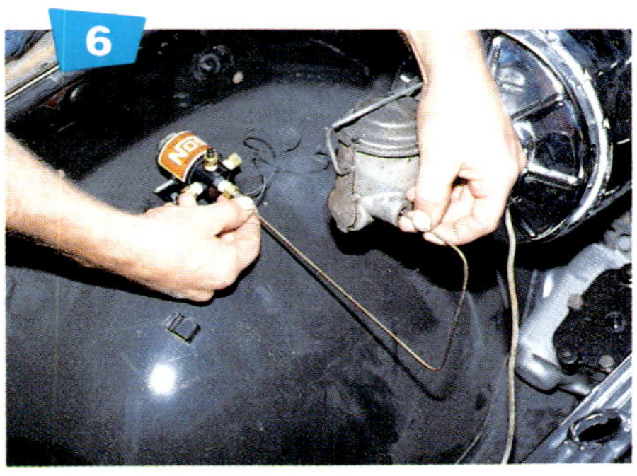

How To FIT A LINE LOCKER

Ever wondered how some cars are able to perform monstrous stationary burnouts?

When you see cars smoking tyres with apparent ease, chances are they're using a line locker. Here's how to fit one in about an hour. Say bye bye to your tyres, though.

A line lock operates by isolating the front brakes from the rear so that they can be applied separately, holding the car still while the driving wheels spin as power is applied. It's basically a high-pressure tap or valve fitted in the middle of your brake lines.

Assuming your car is rear-wheel drive, the line lock needs to be fitted to the front brake lines. When you want to impress your friends with a burnout (or, more importantly, warm the rubber before launching off on a quarter-mile excursion) you press the brake pedal and depress the line-lock button. Your foot can then be lifted from the brake pedal; pressure to the front lines will be maintained while the rears are free to roll. When you want to move forward, open the tap and you're gone. Simple.

The NOS Brake Lock distributed by McDonald Bros Performance in Melbourne is an easy and efficient way of ensuring positive control during the burnout procedure. What's more, by eliminating hydraulic brake pressure to the rear brakes, no undue stress is created. Ever tried performing a big burn in a car equipped with good rear brakes and without a line lock? Unless you plaster the road with water or slippery liquid, chances are you'll make a fool of yourself.

Another alternative, albeit a highly agricultural one, is the old clamp on the rear brake line trick. Of course, this is not recommended, as crimping the line may well result in a fracture and line failure. Read on and you'll see how McDonald Bros Performance fits its $325 NOS Brake Lock kit.

Story and pix:
TREND SETTING PICS

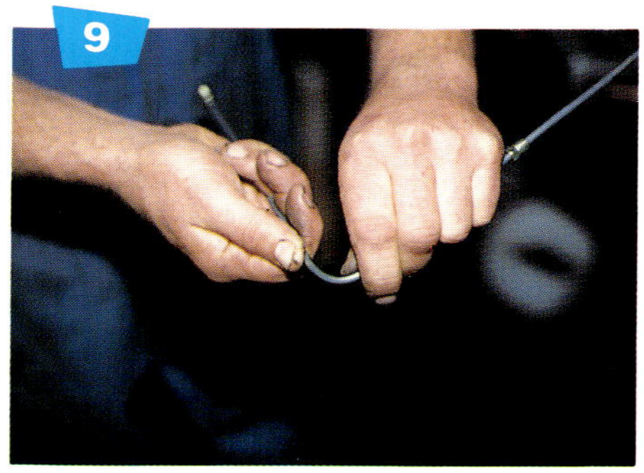

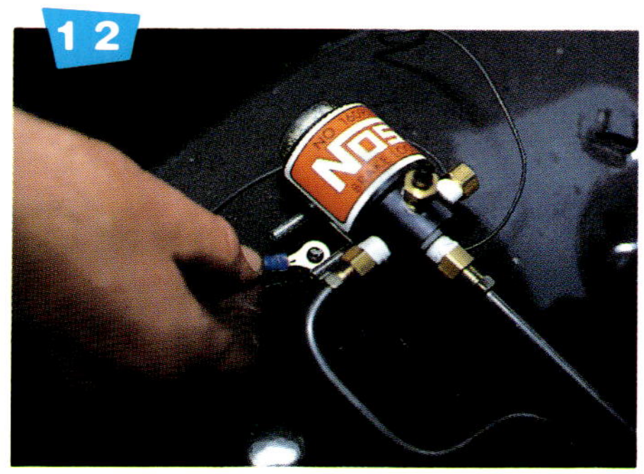

1. NOS Brake Lock pack comes with almost everything needed to complete the job. All you have to provide is some extra brake line.
2. The components removed from the packet ready for installation. The car receiving the kit is a heavily worked HK Monaro complete with tunnel rammed, 400 horsepower, NOS kitted smallblock Chev.
3. Removing the front brake hose from the master cylinder. You must make sure you choose the correct line as master cylinders differ — with some the front reservoir feeds the front brakes, while others have the rear reservoir feeding the front brakes. As a rule of thumb, the larger of the two reservoirs feeds the front brakes.
4. Marking a position for the NOS Brake Lock solenoid. For ease of line fitment and braking integrity, the rebuildable NOS solenoid should be placed as close to the master cylinder as possible.
5. A light covering of Teflon thread sealer should be applied to all fittings.
6. After deciphering which are the inlet and outlet holes (a complete installation booklet is supplied with the unit) you need to design your new brake lines while bearing in mind that the original inlet line may be used if in good condition. This can be done with the likes of welding wire bent to shape. Positioning of the solenoid bleeder valve is important for two reasons: because air rises it should be at the top, and for ease of bleeding it should be placed in an accessible position.
7. McDonald Bros prefers using new line as it bends easily and flares nicely.
8. A double flare. Discern whether

▶ ▶ ▶

How To

the master cylinder requires a single or double flare before carrying out this procedure.

9. New pipe bends easily and smoothly to the desired shape. Cheap flaring tools are readily available and start at around $85. If yours is a one-off situation it's probably easier having your pipe flared by a brake specialist for around $2 per end.

10. Fitting the new inlet brake line.

11. Bleeding air from the solenoid.

12. Wiring is simple in that one wire goes to earth and the other to a fuse. For racecar applications the fuse supplied is recommended.

13. Mounting the Brake Lock button on the steering wheel. However, you may choose to fit it to the shifter knob.

14. Make sure the button is in easy reach of your thumb.

15. Another style of solenoid fitment (this time on a racecar) which is once again placed as close to the master cylinder as possible. Complete the installation by bleeding brakes.

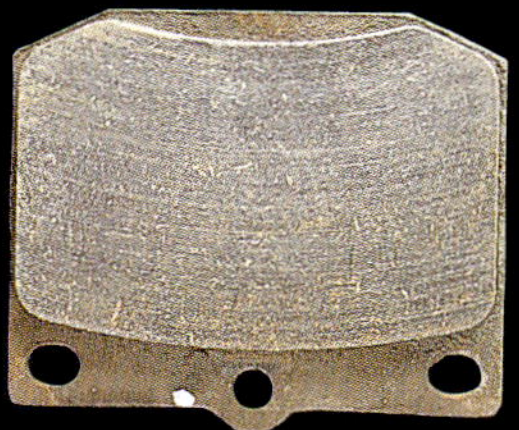

Changing brake pads can make a huge difference to braking performance. You won't be able to change the swept area but you sure as hell will be able to change how effectively the pads grip the disc!

The brakes on a five-litre tourer show what can be done when there's lots of money to throw around. This is one of the back brakes, with a Harrop 4-pot caliper working on a drilled and ventilated disc.

Front brakes on a five-litre tourer show the importance of brake cooling ducts. Note the bell-mouthed entrance, the narrowing of the duct to allow tyre clearance, and the last flexible section. On a hard-driven road car, cooling ducts can also work very well.

RAPID HALT

How to stop better on a budget

Words and pix by Godfrey Towns

Going hard is good fun, no doubt about it. But unless you're into quarter-mile performance and nothing else, it makes sense to be able to stop well, too. Until you're in a car with good brakes you can't really realise the performance advantage that comes from being able to stop so hard that your eyes bulge. As well as impressing your girlfriend (bulging eyes are pretty sexy, yes?) it will mean you're much less likely to get into trouble — the sort of trouble that results in a direct ticket to heaven.

By far the cheapest way of getting better braking is to optimise the brake system you already have. The factory spent good money setting up a system with the right front/rear brake bias, durable and strong components and good parts availability. If your underbonnet power-up is of the 30-40 per cent sort, making the standard system work well is your best bet. Before you even think of spending megabucks on something new, think about what you already have.

PADS

There are pads and then there are pads. The difference between the standard pads (or mebbe the dodgy discount store pads the previous owner put in your car) and hi-po pads is simply immense. High-performance pads will bite hard, bite harder when hot and give stopping distances metres shorter than poor alternatives. The difference really can be that dramatic.

If you're on a budget and drive a relatively common car, look for pads advertised as suitable for towing. I don't care if your car hasn't even got a towbar, a pad designed to haul down maybe three tonnes of boat and car works pretty well with just the car on the road! They'll probably be billed as 'metal' or 'heavy-duty' and come from a reputable manufacturer.

Next up are competition pads. Make sure when you're talking to the retailer or manufacturer you specify that you want the pads for a road car, otherwise the cold braking may be marginal. Look to a form of motorsport where your car is used — HQs, Commodores, rally, circuit — and then talk to the manufacturer who makes pads for them.

Hi-po pads will tend to last a shorter time than factory pads and wear out the discs or drums faster through their more abrasive action. That's the price you pay. Talking about prices, good aftermarket pads are often cheaper than factory items...

BRAKE COOLING

Go for a quick drive and brake heavily a few times. Hop out of the car and feel the temp of the wheels and brakes. Ouch! Hot, aren't they? Brakes lose their effectiveness when they get too hot, so the trick is to keep them cool. Fitting an open-design alloy wheel will let a heap more cool air circulate past the disc or drum, while the alloy will better conduct the heat away.

However, if you're really serious about getting the temps down, brake ducting will do it better than anything else. Adding brake ducts is well within the scope of the average do-it-yourselfer — although still harder than it first appears. It's the front brakes that need to be cooled the most, so you'll be working with disc brakes.

You need to cut a hole in the backing plate behind the disc first, and if it's a round hole that means the whole of the brake disc, caliper, etc needs to come off first. However, you can cheat by bending part of the backing plate away from the disc and then cutting two slots in it. The flap which is then formed can be bent back at 90 degrees to form an attachment bracket, with the backing plate then bent back into its normal position.

So what do you attach to the flap? You'll need to use a flexible tube of preferably at least 5cm diameter. It needs to be flexible because as the wheels turn, so do the brakes! It's easy to forget this...

The clearances between the wheel, tyres, suspension members and bodywork are generally very tight, so measure the gaps before you buy the flexible tube. Even 3cm tube will be a lot better than nothing.

Before the brake duct will flow air it needs to be pressurised by the car's forward movement. This means the mouth of the duct needs to be in clean airflow, and it helps if this opening is larger than the tube itself. The easiest way to do this is to use plastic downpipe and gutter adaptors to give a fairly large rectangular opening that can be positioned under the front bumper or spoiler. You can buy these from your local hardware store and then paint them black with a spraycan.

BRAKE FLUID

Good-quality brand brake fluid has a boiling point of about 290 degrees Celsius. So? Well, you see, brake fluid only works when it stays as a fluid. When it turns into a gas (when it has boiled because of the hot brakes) it becomes compressible, and that means that when you put your foot down nothin' much happens. OK, so when new it won't boil until the temp of the fluid inside the brake cylinder reaches 290 degrees — pretty bloody hot.

Aaah, but with one per cent water, that same fluid's boiling point has dropped to 245 degrees, with two per cent 220 degrees, three per cent 210 degrees, and four per cent 185 degrees! And that's DOT 4 fluid — the best of the lot. DOT 3 fluid with four per cent water has a boiling point down at 150 degrees. And that's the amount of water you'd expect in your brake fluid after just two years!

If you use DOT 4 fluid you should change it every two years, and with DOT 3 fluid have a transfusion every year. There are some super hi-tech brake fluids available, so if you club-race your car, for example, you might want to check out some of these special fluids.

If you're to install brake ducts properly, the brake assembly will need to be dismantled like this until the backing plate is revealed. However, there are some shortcuts that can be taken.

This is the sort of flexible hose that can be used for brake ducts.

RAPID HALT

A ventilated disc (top) will work much better than a solid disc (bottom) because it will stay much cooler. However, unless vented discs and their associated calipers are available cheaply from the wreckers and are a straight bolt-on, swapping to vented discs is out of our budget range.

Drilling discs that were once not drilled is illegal because cracking is almost guaranteed. A shattered disc jamming in the caliper at speed wouldn't be much fun at all...

Brake fluid should generally be changed at least once every two years. It absorbs water through the hoses and its boiling point gets lower and lower as time passes. Good quality fluid is a must.

Do-it-Yourself Safety Check

Out on the streets just cruisin' or trippin' to a major event, our streeters must be in good, roadworthy condition. One of the last things anyone needs is the authorities putting a car off the road for a minor item, and then having to have a wheel change and suspension adjustments to comply with legal requirements.

Breakdowns on long trips are a pain, especially if an inspection beforehand may have revealed any problems needing repair. In following pics are more than 40 items that can be easily checked by the amateur motorist. We can't go into specs on each; we can only guide you on what to look for. The repairs will then have to be completed by yourself or a recognised repairer. If in doubt, leave the repair to the experts. Remember — lives may be at stake.

Raise front wheel and grab wheel diagonal feel for movement. This is to check ball joi tie rod ends. If there is movement check fu by having someone else move the wheel s can see which components have freeplay.

Grab wheel at top and bottom and check for movement. If there's movement, adjust the wheel bearing.

Check hoses for cracks, leaks, perishing or contact with other components.

Remove wheel and check disc pad thickness.

Undo lower bolt on calliper (Commodore calliper shown), loosen top bolt and swing calliper clear of disc for inspection.

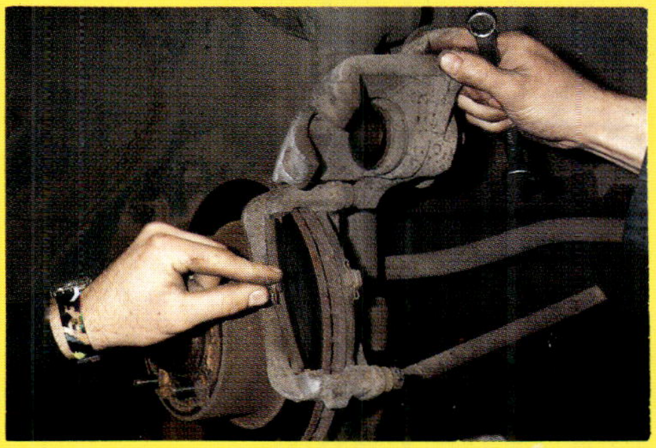

Remove the disc pad for inspection.

Measure thickness of disc and refer to manufacturer's recommendation for minimum thickness.

On left is new disc pad; centre and right are severely worn and need replacing. Excessive wear leads to metal-to-metal contact with discs, causing extensive disc damage and costly replacement.

Remove brake drum and check for brake lining wear, drum scours and grooving.

Worn brake drum shown at front requires machining or replacement.

Do not blow brakes clean with air pressure as the dust is hazardous to health. Use Brakleen spray.

After spraying with Brakleen wipe areas clean with a rag.

Gently pry wheel cylinder dust covers looking for brake fluid leaks. If leaks are evident, have cylinders overhauled or replaced.

▶

HOW TO'S

Excessive wear of brake shoes can result in costly brake drum replacement and unsafe braking.

Serious oil leak at transmission inspection cover indicates faulty engine rear main seal if oil leak present. Or if transmission fluid leak is evident this can indicate a faulty transmission pump seal.

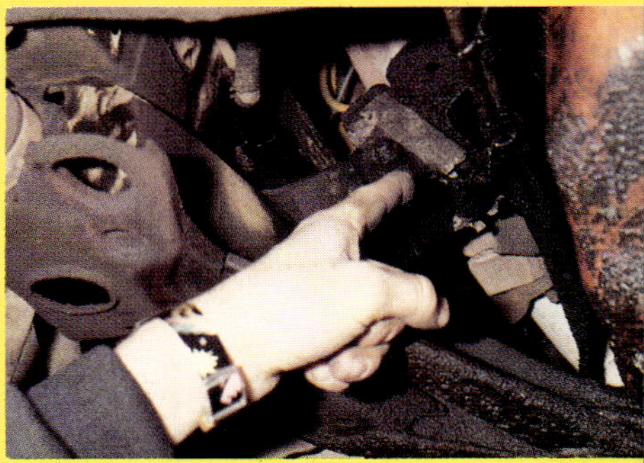

Engine oil leaks can deteriorate engine mounts and rubber bushes. Check also for cracks in rubber.

Check rubbers and bushes on anti-sway bars and suspension parts.

Universal joints: grab both sides and twist in opposite directions looking for movement; replace uni joint if movement present.

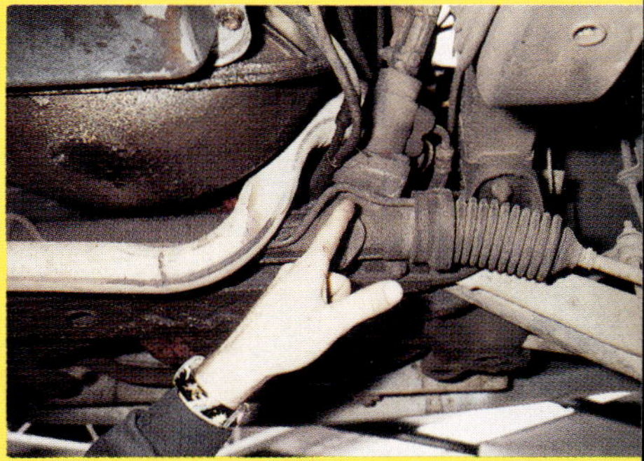

Check for power steering leaks from steering rack and check that dust boots are sound.

Check oil using dipstick and top up if necessary.

Check fan belt tension; allow 1cm up-and-down movement. Too tight a belt can cause wear to bearings in water pump and alternator. Too loose a belt can cause overheating and poor charging of battery.

A perished fan belt should be replaced.

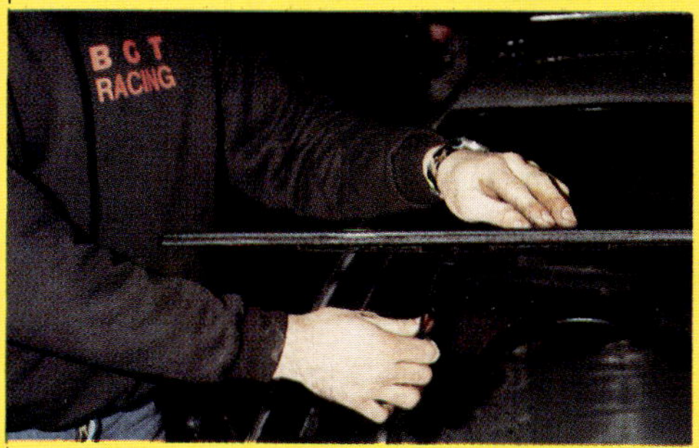

Check wiper blades for signs of perishing; test with water for efficiency.

Check that battery retaining clamp is present and tight.

If residue of rusty water appears, check hoses for leaks or perishing. Tighten or replace if faulty.

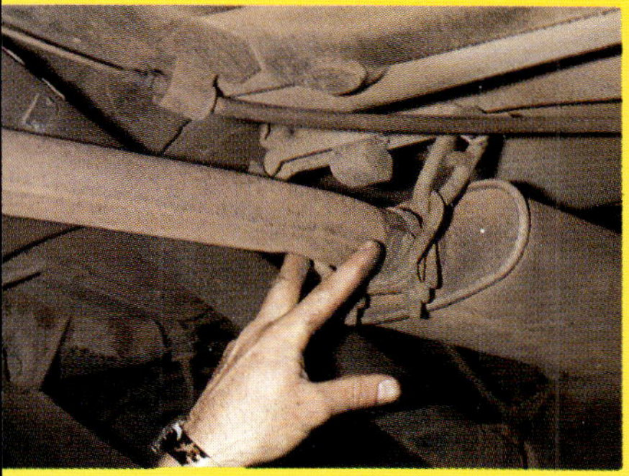

Check exhaust for leaks, rust holes and faulty or missing support O rings.

Check for panhard rod bush wear by checking for movement and shockers for signs of fluid leaks.

Use screwdriver to check for movement in trailing arm bushes. Check rubbers on stabiliser bars.

Check tie rod ends for up and down movement and replace if movement found.

Check brake and clutch master cylinders for leakage and fluid levels. If levels are low, top them up with recommended fluid to marked levels on fluid reservoir.

Check radiator expansion tank and fill to required level. Top up windscreen washer bottle.

Check radiator fluid level.

Check that battery is secure with battery clamp. Clean battery terminals with hot water to dissolve acid build-up then smear with vaseline or grease to prevent acid build-up. Remove cable clamps and clean with round file and clean battery terminals.

Check tyre pressures and inflate to recommended pressure.

Check all lights for operation, alignment and condition.

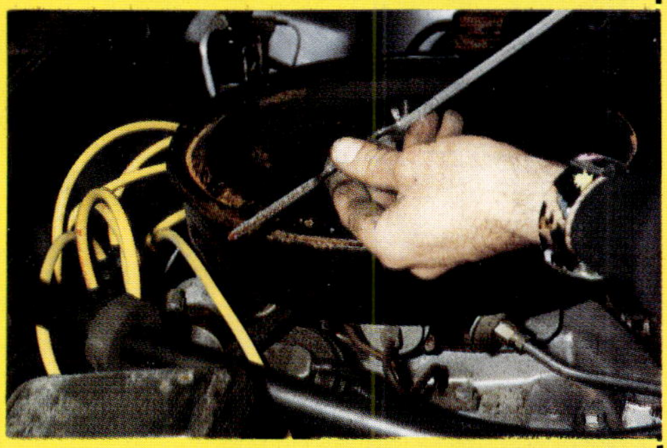

Check transmission fluid while engine is warm and running. If fluid is clean and red it's a sign that the trans is in good order. If trans fluid is dark and smells burnt, the transmission requires servicing or overhauling.

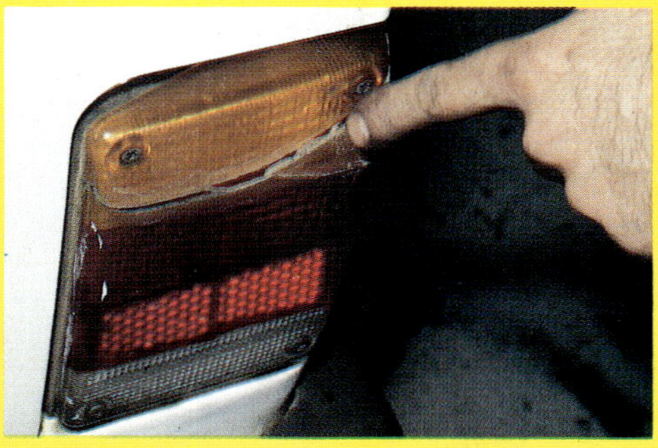

Cracked or faded lenses must be replaced.

Check that horn works!

Check seat belt condition; frayed belts are unsafe and must be replaced.

Check pedal rubber pad condition. Replace if worn or missing.

Check there is sufficient tyre to body clearance; any contact could damage tyres.

Check for uneven tyre wear. Shoulder wear can indicate that suspension or alignment adjustments are necessary.

Detailing on a Budget

Can an old, beatup car be made to look like new?

How many times have you been to a caryard and seen used cars that look brand-new? They might be quite a few years old and have tens of thousands of miles on the clock but still look in terrific condition.

You say to yourself, "Why doesn't my car look that nice after I wash it?"

The answer simply lies in the detail work that goes into cleaning used cars that are to be put on the lot — the finished product must appeal to the eye, hence the expressions 'detailing the car' and 'sending it to the detailer'.

In this story, we bring you a step-by-step guide to doing your own detailing. To find out how to do it the best way, we asked John Towns and his lovely lady, Janet, from Jay-Jay Detailing (018 972 740) to take us through the whole procedure.

We started with a reasonably clean '67 Chevrolet Camaro that is looked after fairly well. It is driven daily and parked on a busy thoroughfare. Although it wasn't too bad, it certainly needed a bit of TLC, especially since the owner was taking it to the Chevy Showdown at Canberra the following weekend.

The whole operation began with degreasing the engine bay after the motor had cooled down. It is not necessary for it to be cold but it should definitely not be hot, as most of the degreaser would evaporate and the car would suffer starting problems due to condensation (also known as perspiration) in the distributor.

All door jambs and boot hinges were also degreased, then hosed off after a few minutes of soaking. The engine bay and engine itself, as well as everywhere else that was sprayed with degreaser, should be sprayed with an industrial cleaner and then washed with soapy water and pressure-cleaned with water. ▶

Story/pix: MICHEL A. MALIK

If your car has an alarm fitted, ensure that it is covered with a plastic bag.

John put the pressure hose to the inner guards from the door jamb and spent over a minute on each while all the leaves, dust and sand where flushed off at the bottom.

The wheels received a few sprays from a mag wheel cleaner and were then washed with soap and rinsed straightaway, otherwise streak marks would have showed.

The body was then washed with car shampoo, dried with a chamois and taken indoors for buffing.

John's been in the detailing business for a long time and watching him perform these routine tasks made it all look easy, although I know there's a fair bit of hard work involved.

After all these years of experience, John knows what products to use and which one will do the job right. Therefore, when I asked him about the buffing compound he'd be using, his answer was, "It's not a standard buffin' compound, but my own mixture."

John used quite a soft combination of polishing compound by mixing Kayes' Calpol Cream and Meguiar's fine cut (No. 2). The polish is not blotted on the car but was spread evenly with a piece of cloth; then a mist of water was sprayed on the car. That, I'm told, stops you from inadvertently burning the paint while you buff it.

The hard bit done, John went on to hand-polish the car using Meguiar's Final Inspection (No. 34). It should be spread evenly, panel by panel, using a soft cloth, and then simply wiped off as if you were polishing by hand — but without the swirling motion and the hard rub!

The body done, John's attention turned to the tyres which he painted with flat black tyre-black and then sprayed with a vinyl restorer, such as Armorall. Janet said they do not always spray the tyres with vinyl restorer because some customers like the flat, dull look on tyres.

The windows were next in line and were wiped with Amway's Crystal Clear Concentrate. John's advice here was to always clean the wiper blades as they can make a mess if they are overlooked.

Janet mentioned a trick they use to remove overspray or dried dead bugs: humble steel wool when used wet poses no threats as far as scratching the glass is concerned.

The next phase of the detailing was the interior, which was Janet's specialty. Basically, the interior was vacuumed completely using a large vacuum cleaner and a small brush for the hard-to-get-at places.

The dashboard, armrests and rear parcel shelf, as well as the console, gear lever, indicator and wiper stalks, and steering wheel were then cleaned with Rite Off (an industrial-strength upholstery cleaner) and finally dressed with vinyl restorer.

The glass was wiped from the interior while the door jambs, under-boot and under-bonnet were thoroughly checked for any trace of polish which, if found, was removed.

Finally, while Janet dressed the bumper bars with vinyl restorer, John filled his spraygun with GMH black and coated the inner wheel arches as well as the lower rear (under the rear bumper, basically, on the undercarriage) and lower front of the Camaro.

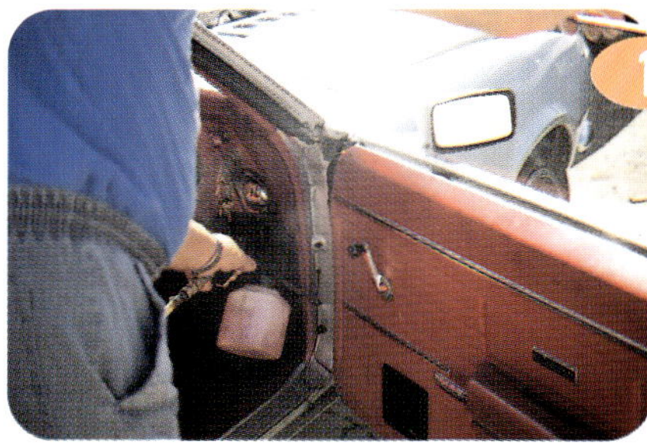

1. The door jambs are degreased.

2. The inner-door edges are also sprayed with degreaser.

3. The wheels receive a squirt of mag wheel cleaner.

4. The wheels are soaped straight after the mag wheel cleaner, as the paint could be affected if the cleaner is left on for too long.

5. A hose-off with a high-pressure hose finishes the wheel treatment.

6. High-pressure water is used to clean the outer edge of the door.

7. Door jambs are treated with high-pressure water as well.

7

8

9

10

11

12

13

14

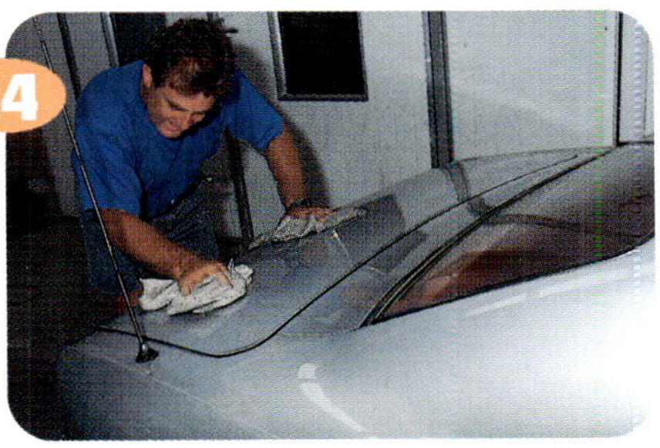

15

16

17

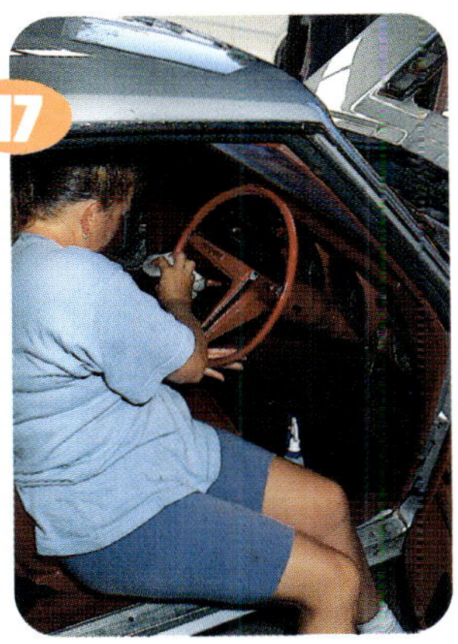

18

19

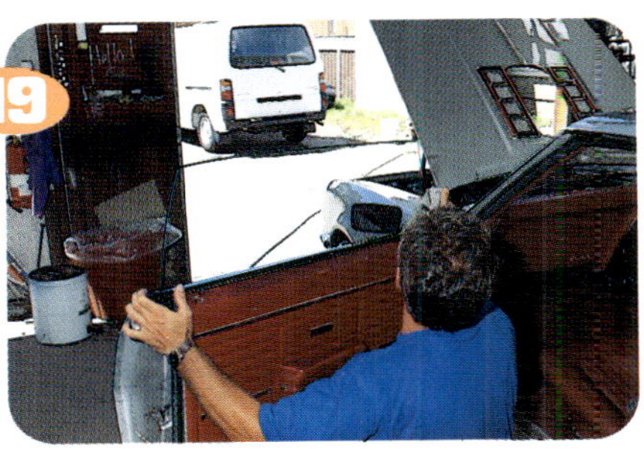

8. An industrial cleaner is sprayed after the degreaser is blasted off with water.

9. Just as the wheels are treated, the door edges get a workover with soap to neutralise the industrial cleaner's powers.

10. Inside the workshop John wipes on the buffing compound with a cloth.

11. A squirt of water on the buffing compound stops the paint burning.

12. Buffer firmly in hand, John goes over the Chevy thoroughly until the shine comes back to it.

13. Meguiar's Final Inspection (N034) polish is sprayed on.

14. John hand-polishes the Camaro to a gentle and soft finish.

15. The wiper blades receive as much care as the glass they are meant to keep clean.

16. The tyres are sprayed with a vinyl restorer to give them extra protection and lustre.

17. Like the rest of the interior, the steering wheel is wiped and wiped and wiped . . . till it's absolutely clean.

18. The doortrims receive thorough treatment; after the clean-up they are wiped with vinyl protector.

19. Final touch: the mirrors get the treatment with a Meguiar's Final Inspection.

FIRES IN CARS

They do this all the time at the training college.

After spending years of late nights out in the garage working on your pride and joy, you definitely don't want it to end in a spectacular fireball that could have been avoided.

We got to thinking about cars and what makes them burn. The obvious organisation to talk to about this is the fire department so we went to the MFB. Being the efficient organisation that they are, they had prepared the following tables of information for us.

The numbers shown here make an interesting study. Insufficient data generally means that the car was burned to the ground and there was no trace of the cause remaining.

External areas usually involve suspicious circumstances because it's not very often that the paint on a car spontaneously ignites; it usually has help.

TABLE OF DATA FOR FIRES IN CARS IN MELBOURNE DURING 1996

AREA OF FIRE	PART FAILURE, LEAK OR BREAK	SHORT CIRCUIT, GROUND FAULT	OTHER ELECTRICAL FAILURE	BACKFIRE	MECHANICAL FAILURE	EXHAUST SYSTEM	TOTAL
Insufficient data	0	0	0	1	0	0	1
Passenger area	0	19	6	0	1	1	27
Luggage	0	0	2	0	0	1	3
Engine area	90	58	21	50	4	14	237
Fuel tank or fuel liners	6	0	0	1	0	2	9
Driver's area	0	6	0	0	1	0	7
Exterior exposed	2	0	0	0	0	1	3
Totals	**98**	**83**	**29**	**52**	**6**	**19**	**287**

Not surprisingly, most fires occur in the engine area. The next most common location is the passenger area. Often, the cause can be something as simple as a cigarette butt carelessly discarded, but electrical faults also feature highly in this area.

TABLE OF DATA FOR FIRES IN CARS IN MELBOURNE DURING 1997

AREA OF FIRE	PART FAILURE, LEAK OR BREAK	SHORT CIRCUIT, GROUND FAULT	OTHER ELECTRICAL FAILURE	BACKFIRE	MECHANICAL FAILURE	EXHAUST SYSTEM	TOTAL
Insufficient data	0	0	0	0	0	1	1
Passenger area	7	28	5	0	1	3	44
Luggage	0	5	1	0	0	2	8
Engine area	146	81	27	98	8	20	380
Fuel tank or fuel liners	6	1	0	2	1	2	12
Drivers area	0	2	1	0	0	0	3
Exterior exposed	0	0	0	0	0	1	1
Totals	**159**	**117**	**34**	**101**	**10**	**29**	**449**

Speaking of fires in the passenger compartment reminds me of a story a friend told me. Apparently, some guy had run out of petrol and some mates in another car had given him a lift to the petrol station. They couldn't rustle up an appropriate container so they'd filled some sort of open dish. As they were all crammed in the car on the way back to the stranded car someone in the back lit a cigarette (makes you wonder, doesn't it) and the dish of petrol caught fire. The guy holding it in the front went to throw it out the window in a panic but the window was shut. You can guess the rest. If you have mates like this - get rid of them. Make sure that you only ever transport fuel in appropriate containers.

Even things as seemingly innocent as linseed oil on a rag can cause a fire. In this case the oil would decompose and give off heat in doing so. If the heat is trapped because the rag is bundled up fairly firmly, it can feed back on itself creating a constantly rising temperature with fully fledged combustion as the result. Be careful next time you're at cricket practice!

This inability for heat to dissipate is the cause of fire in many circumstances. Removing the source of heat is one of the three keys to controlling a fire so anything that prevents heat loss is a potential source for creating a fire. As the fire department pointed out, a cigarette butt smouldering on the seat of an office chair probably won't start a fire. If, however, it gets stuck in a crevice where the heat it produces remains trapped, then the potential for fire is greatly increased. In a modern car this is a complete disaster because of the amount of plastic interiors are trimmed with these days. Really old cars were trimmed with things like real leather and kapok so interiors didn't burn as easily. Dash boards tended to be metal so they didn't burn anywhere near as readily as current versions either.

We went down to the MFB/CFA training ground to see the proper method of putting out a fire.

It's best to avoid feeding the oxygen to the fire so just open the bonnet enough to get the nozzle under it.

Discharge your extinguisher until the fire is out or you run out of powder.

Here's what can happen if you do fling the bonnet open.

PREVENTION

Servicing your car regularly is one important step toward preventing car fires. Checking that things like fuel lines and other fuel-related components and wiring systems remain in good order goes some way to preventing problems. We're all fairly careful about our project cars but what about the parts runner? These should also be checked. The fire department also said that problems can occur after accidents because people don't always check to see how wiring has been affected by crumpled bodywork. This situation can create short circuits in otherwise perfect wiring systems.

LPG is a perfectly safe automotive fuel just like petrol if it's installed correctly. These days, with everyone working on LPG being licensed, installations tend to be of an adequate standard but this wasn't always the case. Cars that were fitted before the regulations came into effect are still legal but some of the methods used can leave something to be desired. Retaining straps welded to tanks, plain copper supply lines with no external protection, no flexure built into the lines and other things can all create problems.

If you are buying an older car fitted with gas, check the condition of these items. Make sure that the cylinder is vented properly so that gas can't accumulate in

Obviously, we burnt this car on purpose.

Clearly, we've helped this fire along for the purpose of photography, but it's a reasonable simulation of what can happen.

If unchecked, the fire will soon reach the interior.

the boot. If it does, just putting your foot on the brake could cause a fire as the brake lights are switched on.

Static electricity can also cause problems. This is why petrol tankers earth themselves as they're delivering petrol. When we fill our cars we are earthed through the supply gun and line which has special provision for dissipating static charges. Even fuel moving through the hose can create a static charge which has to be dissipated.

Inspect the engine bay yourself. If you smell fuel when you shouldn't, stop the car, turn off the ignition and investigate. The smell of smoke should generate the same reaction with the additional step of calling the fire brigade.

FIRE FIGHTING

A fire needs three things to sustain itself: fuel, oxygen and heat. If you remove any of these three factors the fire will go out. Extinguishers smother the fire and prevent oxygen from reaching the fuel and can also have a cooling effect depending on the type. If you do detect a fire and it hasn't progressed too far, you can try to put it out yourself if you have an extinguisher although you should still give the fire department a call.

Yellow BCF extinguishers are still around but it is now illegal to discharge them because of the propellant. They were an excellent type of extinguisher but they wreck the upper

It only takes about 60 to 90 seconds to get a fire truck on the road, but by the time the call comes in and they drive to the scene, this is the stage your average car reaches.

The fire department is obviously good at putting out fires but...

If they do have to come and put it out for you, this is more than likely what you'll be left with.

atmosphere. If you have one of these lying about you should take it to your local fire station for disposal.

CO_2 also works very well, but being a gas, a fair proportion of it can be blown away on a windy day. Also, most CO_2 extinguishers are reasonably large and not suited to being carried about in a car. They are, however, a good choice for garages and can be used against flammable liquid and electrical fires.

Dry Powder extinguishers come in compact little sizes that are well suited to being tucked under the front of the driver's seat or any other handy location that doesn't get in the way. They are suited to all types of fires that you'll generally encounter. Really, these are the choice for cars.

Blue foam-type extinguishers are also excellent but they're only suited to garages because of their size and shouldn't be used on electrical fires.

Whatever type of extinguisher you have, make sure you are completely familiar with how to use it. It will cost precious time if you have to fumble about trying to work it all out while a fire is actually burning.

INSURANCE

Fires in cars that are registered and running are bad enough. But what about when you're building it? At that stage it won't be covered by insurance unless you make special arrangements. Even though your project may be incomplete, you'll still have quite a bit of time and

FIRES IN CARS

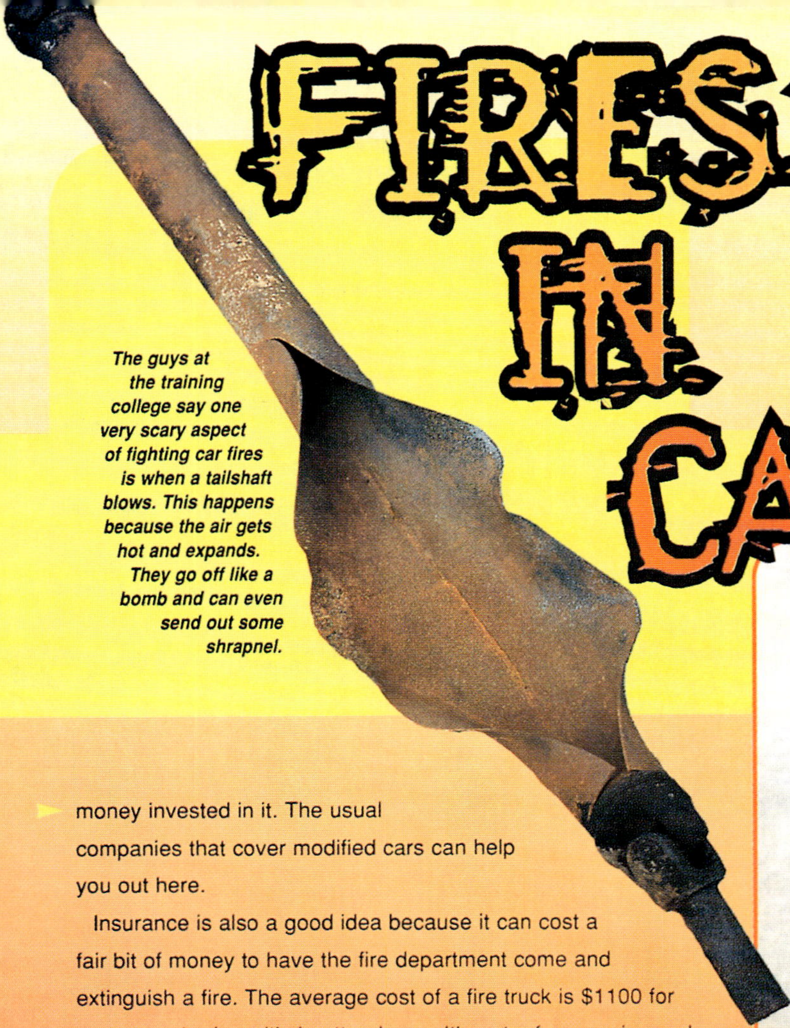

The guys at the training college say one very scary aspect of fighting car fires is when a tailshaft blows. This happens because the air gets hot and expands. They go off like a bomb and can even send out some shrapnel.

CO2 extinguisher.

Foam extinguisher.

money invested in it. The usual companies that cover modified cars can help you out here.

Insurance is also a good idea because it can cost a fair bit of money to have the fire department come and extinguish a fire. The average cost of a fire truck is $1100 for every quarter hour it's in attendance. It's not a free service and it's not entirely paid for by the government like you'd think. In Melbourne 75% of the service is paid for by insurance companies with the state government making up a further 12.5% and local government contributing the remaining 12.5%. The funding structure differs around the country but it's generally similar to this. Fortunately, however, the MFB don't charge for car fires but we're not sure if it's the same story around the rest of the country. Ring your local fire department and find out their policy on the matter.

So, a great deal of fire fighting rests in preventing fires before they get started. Check everything on your car regularly, keep it in good condition and you shouldn't have too many problems. If the worst does happen, stay calm, get your extinguisher (you will have one after this, won't you) and follow the steps outlined in the article. Here's hoping you'll never need to use it!

A selection of dry powder extinguishers. The smaller versions are best suited to cars.

STORY/PHOTOS BY GREG BRINDLEY

ENGINE FATIGUE

YOU'VE PAID STACKS for a new engine only to have it fail after a couple of thousand clicks and, to say the least, you're pissed off. And rightly so. Mind you, though, it's not always the fault of the engine builder. But how do you determine who or what is at fault?

As this is a performance-oriented magazine it's logical that your immediate reaction relate to high-performance rebuilds and the like. Thing is, reco engines far outweigh performance combinations; for that matter, engine or any other segment of driveline failure isn't restricted to performance cars. The typical engine failures that normally occur are not due to extracting extra power; rather they're more likely to be the result of incorrect machining or assembly of the engine.

THANKS FOR YOUR HELP

Determining what caused minor to catastrophic damage in an engine is where the bone of contention normally lies, of course, with the average engine builder claiming to know a lot more than the consumer. Hence we're somewhat at their mercy.

Microscope is often used to locate problem areas.

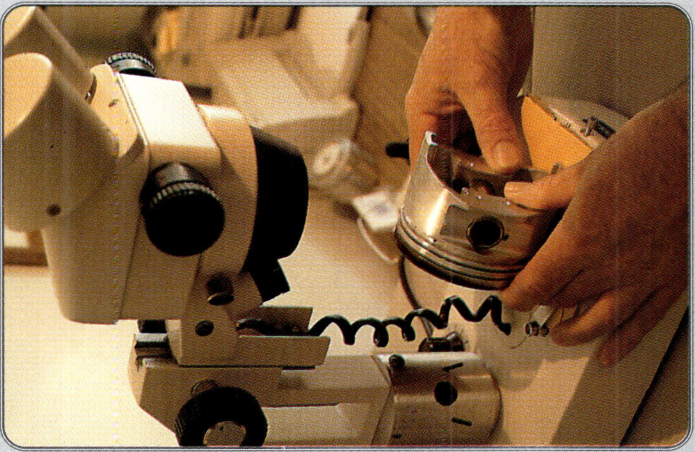

Inspecting ring.

But this need not be. In cases where a problem results in a dispute, the most logical recourse is to call on the likes of Ron Murphy of Australian Automotive Solutions.

Ron is one of an elite few specialising in deciphering the cause of damage to all types of driveline members. For example, it's common knowledge that while the engine is by far the most problematic, the clutch, transmission (manual or automatic) and differential can all suffer major failure. Layman inspection of the damage often isn't enough, and that's when Ron comes into the equation.

The same applies with a professional race car team where, during the course of normal development, breakages occur. In any case, Ron is able to offer concrete analysis which may allow the engine developer to incorporate modifications to prevent future problems, which may, in fact, place them in front of the opposition.

Custom-made ring load tester.

Ron's expertise is the result of years of involvement in the design and manufacture of all types of motor vehicle components in his role as a mechanical engineer. Thanks to the retirement of a close friend who had been running a successful analysis business for some time, Ron moved away from his position with Repco to take up the sometimes-maligned position of providing automotive solutions. Placing himself in the battlefield means he's bound to be hit with some flak, but when it comes to litigation there's normally a winner and a loser and Ron is the one responsible for providing information that helps with the decision.

HOW IS IT DONE?

It's impossible for us to even scratch the surface of the intricacies in just a few hundred words, so we'll simply look at the type of equipment involved.

HARDNESS TESTER: As the name implies, the hardness tester is used to measure the hardness of various engine components including crankshafts, camshafts, valve lifters, valves and pistons. To establish the condition of these components, the measurements taken are compared with the manufacturer's original specifications. In the case of pistons, the hardness tests are also a way of determining whether the engine has been exposed to either detonation or pre-ignition.

MICROSCOPE: The microscope is an invaluable tool in determining any manner of engine problems. It gives the investigator the ability to scan components in minute detail for such things as cracks, wear patterns and foreign particles (dirt).

RING LOAD TESTER: This is used to measure the tangential loads exerted by the piston oil ring assemblies. This value can then be compared with manufacturer specification plus it can be used to calculate the actual wall pressure exerted on the cylinder bore by the contact faces of the oil ring.

Is there an independent motor assessor you can pay to determine why your newly built engine just blew up? Sure is. Read on and find out why an independent assessment is your best choice.

CHEMICAL TESTING: Chemicals such as potassium ferricyanate are used to determine the presence of metal particles embedded in components such as pistons and bearings. Copper sulphate is an accurate measure to establish the condition of chrome-plated surfaces such as cylinder bores and piston rings.

RIGHT & ABOVE: Piston is exceptionally badly damaged, thanks to detonation.

VOLUME MEASUREMENT: Metering equipment is used to accurately measure the volume of cylinder heads, piston crowns and cylinder head gaskets in order to calculate compression ratios.

DIMENSIONAL EQUIPMENT: A large range of linear measuring equipment is necessary to accurately inspect any machined surface in an engine under investigation.

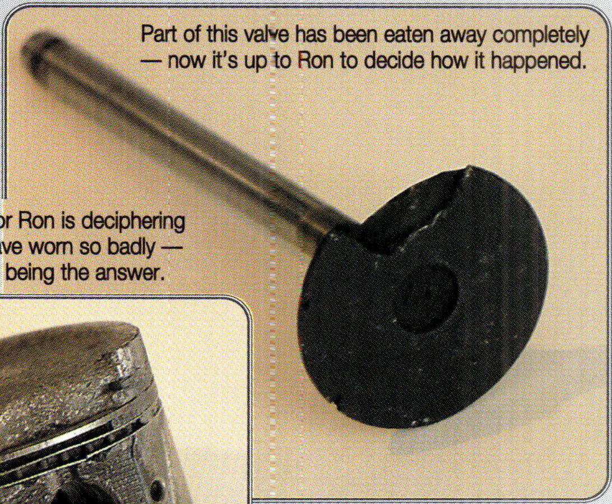
Part of this valve has been eaten away completely — now it's up to Ron to decide how it happened.

BELOW: Another job for Ron is deciphering how these bearings have worn so badly — with foreign particles being the answer.

TESTS PERFORMED

Just as an engine builder is normally able to assess how long-term wear has come about via visual inspection, Ron often bases his test procedures on visual results. His vast experience comes to the fore here with knowledge of what to look for when given a set of circumstances that have led to engine failure.

It doesn't always work, of course, but the benefit is that the number of tests can often be reduced, hence a lower cost to the customer.

A result is achieved in all cases and passed on to the customer. "From my visual observations and dimensional measurements, I write a professional report in a language that's easy for the non-professional to understand," he explained.

Ron went on to explain how much of his work comes about. "Reports are requested for many reasons. It could be that the engine has failed shortly after being worked on and therefore, before any additional work is carried out, the cause of the problem must be established. It could also be an insurance or warranty claim that requires rectification before any assessment for damages can be resolved.

"A dispute between various parties can also be the reason for these reports and in these cases it can either be resolved once the report is finished or, alternatively, it may go to court, in which case the report is used and if necessary supported by myself as a professional witness."

BELOW: Gudgeon pin wear is a result of improper parts-cleaning during reassembly. Grit on gudgeon pin bores quickly became embedded, resulting in a sandpaper effect.

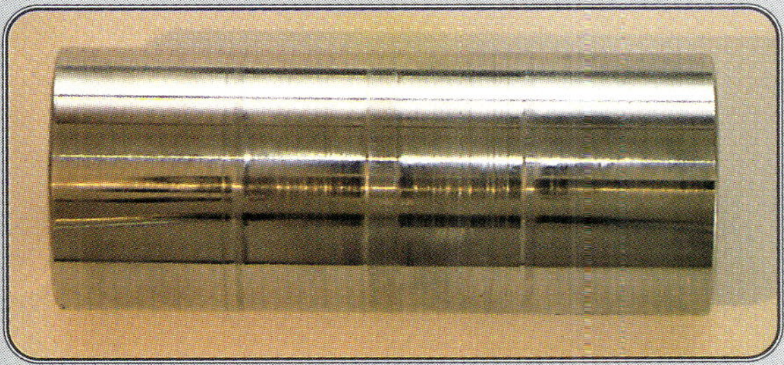

RIGHT: Although minor when compared with the damage often caused by detonation, because the ring land broke, it made the engine unserviceable.

WHY BOTHER?

Personally, I have faced the proposition of having to fork out a not-inconsiderable amount of money on rebuilding an engine that had only just been rebuilt. What was I to do? Pay the same engine company to built another grenade, go with my tail between my legs to another engine builder or talk to someone like Ron who has the ability to put up an accurate answer to my dilemma?

I won't tell you what I did do. Suffice to say that, rather than spend another $5000, a favourable report — one indicating negligence on the part of the builder, or perhaps a faulty product — would mean that 1) the problem must be rectified at the engine builder's expense and 2) out-of-pocket expenses, such as the cost of the report, must also be paid by the engine builder.

CONTACT:
Ron Murphy. Australian Automotive Solutions
Christmas Hills, Victoria   Phone/Fax: (03) 9719 7659

Terminal crankshaft.

Note the 'polished' centre cylinder head chamber — the result of a small component making its way from the carby into the engine and damaging it.

How To fit door & ignition locks

There will always be some jerk out there who would rather nick your car instead of paying for his own

LOCK IT UP

The last thing we need in these trying times is to be forced to repair damage to our cars caused by thieves and vandals.

If the locks are damaged, the average bloke will either strip a replacement off a spare car door he's got at home or head for the wreckers.

New is better. Sure, there's a greater initial cost, but the added security of new locks will reduce the risk of future theft. There'll be less chance of lock jiggling or picking by an amateur locksmith (read casual thief or joyrider).

LOCK MECHANICS

The average car door lock consists of a tumbler or leaf arrangement which lines up with the humps and bumps of your key. Once these are aligned correctly the mechanical part of the lock assembly can operate and unlock the door latches or work your ignition switch.

Locksmiths are trained in lock jiggling or picking as part of their profession. It's a real art. You may think, as I did, that you're pretty clever when you can pick a lock or use somebody else's key to open a different lock. But sorry, we ain't that clever after all; the average bloke can do this only when the lock housing is faulty or worn.

The components inside locks are made of steel and brass, with a few extra spring-steel bits thrown in for good measure. These bits have been refined in design over the years to provide reliable service.

The housings or main bodies of car locks are produced from diecast alloy. It's a cost-effective and reasonably adequate method of manufacture which should offer acceptable service life. The trouble is, when we put our keys into the locks of our cars we immediately start turning them. We should push the key all the way into the lock and *then* start turning.

The fact that we don't puts excessive strain on all parts of the lock and accelerates the wear on the diecast lock housings. It's this wear that enables the lock to be jiggled or picked easily.

Story/photos: JAMES GARROD

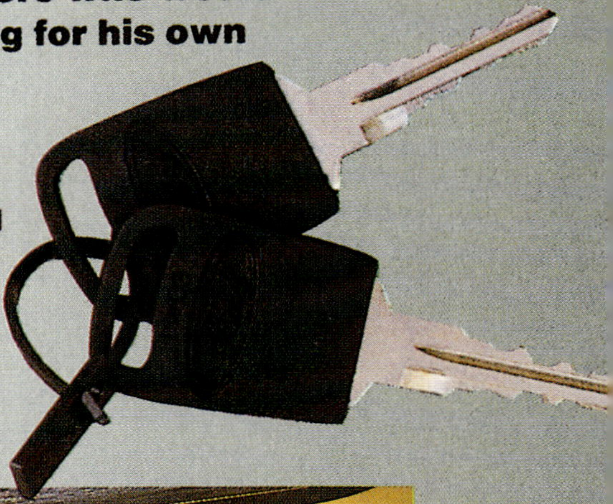

1. Butchered door lock — at least they didn't dent the panel!

2. Equally butchered ignition barrel and surround. This time it cost us heaps in parts and labour.

The case study for this story is an HZ Holden, one of the easiest cars to pinch of all time. It gets left outside a tech college all day and makes a simple, attractive target for a thief — V8, mags, stereo, decent seats and all.

The door lock was forced and the alloy steering column surround/ lock-housing severely damaged by inexperienced thieves. They didn't get away with the car but the damage they caused was costly to rectify.

NEW NOT OLD!

New door and ignition locks are still available from Holden dealers. The locks and procedures described are similar, if not identical, for many different models of Holdens from HQ to WB, Toranas from LJ to UC, and some Commodores.

Now here's the clue: if you ask for two door locks and an ignition lock assembly, the chances are you'll get what you ask for, along with three different keys to add to your keyring!

If you're clever, you'll ask for a complete ignition lock and keys, and a pair of uncoded, new door locks. These bits should cost you around the hundred dollar mark but you'll then have to visit a locksmith to get the door locks keyed to the new ignition barrel. This shouldn't cost more than around forty bucks and you'll end up with only one key plus the knowledge that it's all brand new.

We bothered WH Raisbeck & Sons (Locksmiths), who had already helped us out with a previous problem, to give us the technical info and rekey our new locks for this story. Gerard or John can rekey, rebuild or repair and supply and/or fit just about any lock from a car to a bank vault. They're both practising street machiners and will take good care of your machine. Their numbers: (02) 899 5568 or 899 5569.

FITTING THE IGNITION LOCK

We did the dirty bits (the fitting) ourselves. Anybody with a bit of brains and a tool kit can do the job easily — even us!

3. The steering wheel must be removed. Undo the centre nut and pull towards you with even pressure using hands and knees. If it won't come off, use a puller — but never a hammer or screwdrivers to lever it!

4. The steering column must be dropped to enable easy access to the column and inner workings and wiring. Three 9/16" bolts are all that hold it up.

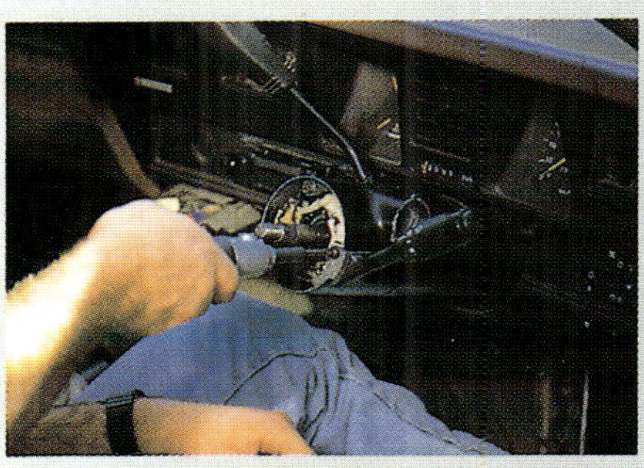

5. One screw holds the indicator arm/stalk in place and three or four hold the indicator mechanism. Remove them carefully so they don't drop down inside the column. A good tip here is to wipe an old speaker magnet over your screwdriver, which will magnetise it just enough to hold the screws.

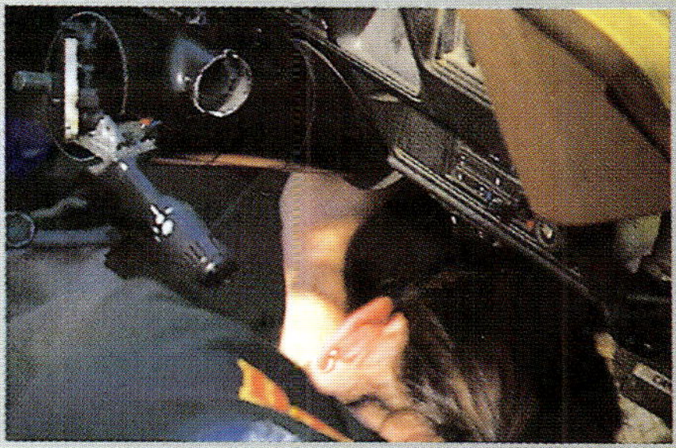

6. Disconnect the two wiring harnesses for the indicators and wiper controls (if so fitted) from under the dash. You should be able to carefully pull the wiring up through the steering column and remove them completely.

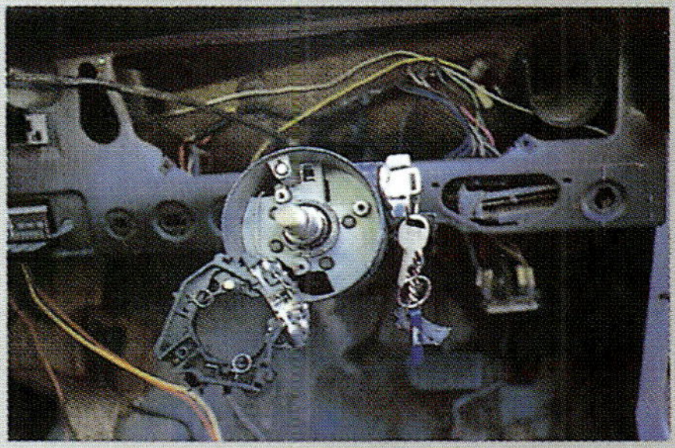

7. If the ignition lock barrel is still in the surround, you'll need to insert the key, turn it to the ON position and insert a thin screwdriver into the slot located near the column-end of the lock housing. If you can't find the slot, don't worry, it's there but may be covered by casting flash. Some probing with the screwdriver will break through this flash and reveal the slot.

8. Four small bolts will need to be removed from the top shroud of the column. These are self-tapping type bolts and can break easily, so be careful! Removal of the circlip, washer and spring from the steering column should also be performed at this time. This car did not have them fitted — someone had obviously lost them — but they're an integral part of the steering as they pre-tension the top shaft bearing.

9. Pre-painted replacement shroud/lock support is then re-fitted, paying careful attention to the correct fitment of the lock operation rack gear and the steering wheel locking pin. It's easier to remove the nylon gear from inside the shroud assembly with long-nose pliers, hold the rack gear up with a screwdriver and then refit the nylon drive gear. Check its operation with a screwdriver to ensure full ignition operation (including accessories switch operation) before inserting your new lock barrel.

10. With the key turned to the ON position, you should be able to push the new lock into place and feel a distinct click as the spring-loaded retainer locks the assembly into place.

Reassemble the steering column in the reverse order to the disassembly and, barring further attacks with slide hammers, you should have years of trouble-free locking, operation and increased security.

DOOR LOCKS: YOUR FIRST LINE OF DEFENCE

Door locks offer very little defence against a dedicated crim. Even if they were bullet proof the thief would just bust a window and get in.

To slow down the jigglers and pickers, though, you should replace damaged locks with new items, as mentioned earlier.

In early-model Holdens as well as some other makes, you could just pull on a small metal tab inside the door frame to remove the locks.

In the HQ to WB Holden, later Toranas and Commodores, the lock retention method is the same but the door trim must be removed to change the locks. Follow these steps to learn how easy (or hard) it can be.

11. With a large Philips-head screwdriver, remove the door armrests. This will reveal, depending on the model of car, another screw that retains the door opening lever. If you have stereo speakers fitted, these should also be removed at this time.

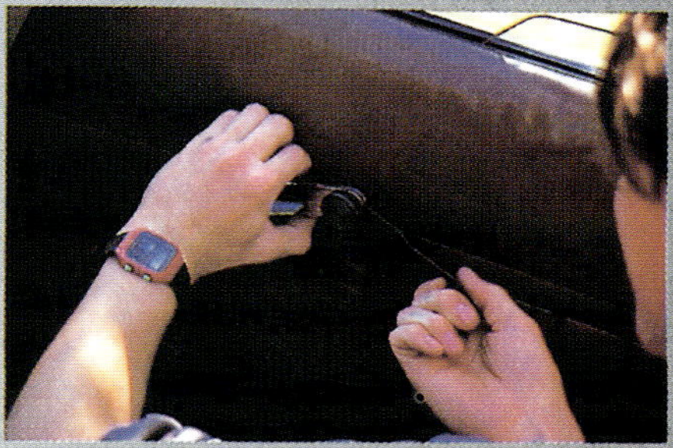

12. With a winder-handle-clip removal tool (available for a few bucks from any auto shop), remove the winders and/or door handles if these are retained by clips.

13. The same tool can also be used to gently prise the door trims from their retaining clip holes in the door frames. Use of bare hands or a screwdriver can damage the door trims. Whilst some trims will fall straight off, HQ to WB types need to be swung outward from the bottom and lifted upwards to unclip from the window channel opening.

14. After making sure the window is wound all the way up (you may need to refit the winder to do this), feel inside the door around the lock.

You'll feel a spring steel plate with a 90-degree bend facing towards you. This plate has to be pulled away from the lock body in parallel to the inner door skin. If it doesn't move freely by hand you can use pliers or multigrips.

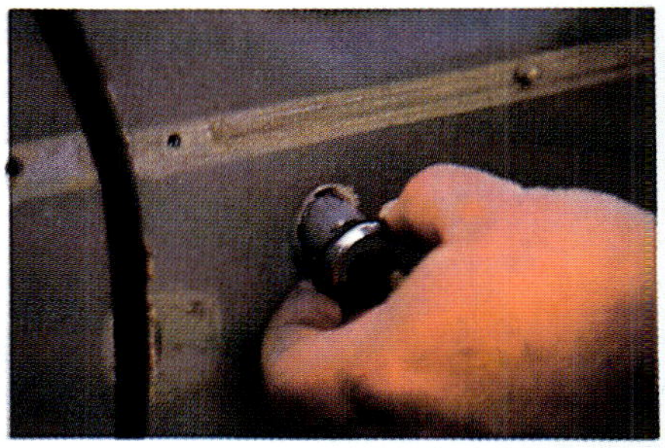

15. With this spring plate removed, the lock assembly should become loose at the outer skin of the door. Move outside the vehicle and carefully move the lock around until most of it is outside the door. An angled steel bracket is connected by a wire linkage to the door latch assembly. It should disconnect easily, with a bit of twisting.

16. Take care not to damage the small rubber gasket/seal around the old lock when you remove it. These are not supplied with replacement locks, although they should be, and are necessary for correct fitting tension and water sealing.

To fit the new locks, reverse the removal procedure, taking care not to damage paint or panels in the process.

We strongly recommend the fitment and use of a quality alarm system or some form of extra security device, like kill switches, fuel cutoffs, wheel clamps, locking nuts and the like. The more anti-theft devices you have, the less likely you'll be to lose all or part of your dream machine.

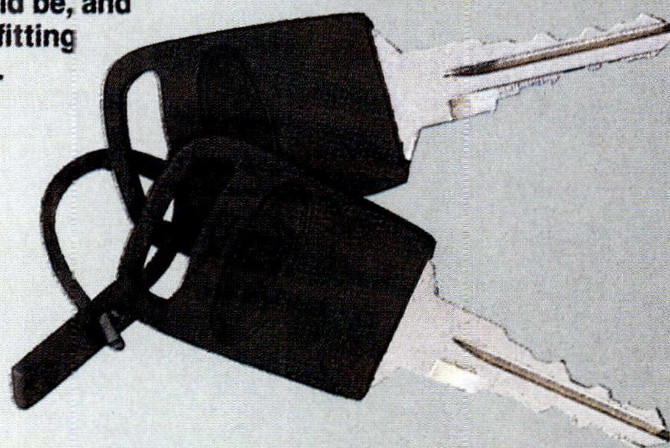

SECURE YOUR DREAM

Ever had an ignition switch torn out of your Ford with a slide hammer? I have. But it won't happen again!

story/photos JAMES GARROD

No matter what you do to prevent your machine from being nicked, if the thieves really want it, they'll take it. There are things you can do, however, to make their job much harder.

Alarm systems can be effective; kill switches are a damn good deterrent, too; plus, I have a hardened steel cable and padlock wrapped around my car's steering wheel as well.

If you slow thieves down as much as possible they are likely to give up... and steal someone else's car instead. I strongly recommend using one or all of the above theft-deterrent measures, but just check out what else I've fitted to my latest V8 toy.

Shielding your barrel

Falcon ignition-lock barrels are notoriously easy to pick as they get on in years. Even if the thief doesn't know how to pick the lock, he usually just slide-hammers it out and you've lost your car. The solution to the lock-picking problem is to fit a better lock barrel. In the case of Fords in the XC to XF range, there is a substitute lock barrel available from a different model which is more resistant to picking.

This does not remove the slide hammer threat, though, but there's a shield/shroud arrangement available which can drastically reduce the chances of success by this method. The Squire Security Systems Ignition Lock Shield is a well engineered kit specifically designed to reduce lock picking and slide hammer removal.

The principle behind the system is to replace the ignition lock barrel with a brand new, modified Cortina

1 This is what your ignition lock looks like after it has been ripped out with a slide hammer. It's really easy to start the car and drive away once the lock is gone! Fortunately, the column surround was not damaged by the would-be thieves or the repair bill would be much higher.

2 On XC to XE model Fords the steering column needs to be dropped to gain access to the column shroud and lock assembly. Three half-inch or 13mm nuts hold the column up, and when removed allow the column to drop just enough to remove the Philips-head screws that retain the shroud. XF models may not need the column to be dropped as the shroud screws are readily accessible.

3 This is the complete kit, as supplied, containing new lock cylinder, two keys, steel surround, hardened steel end plate, break-off retaining bolts, lubricant, instructions and warning stickers. The end plate covers the area where the key is inserted and freewheels when no key is in place.

4 Even the experts know that instruction sheets are there to be read and followed. You may think the procedure seems self-explanatory — BUT — there is a strict order to follow and you may stuff up if it isn't followed to the letter.

2

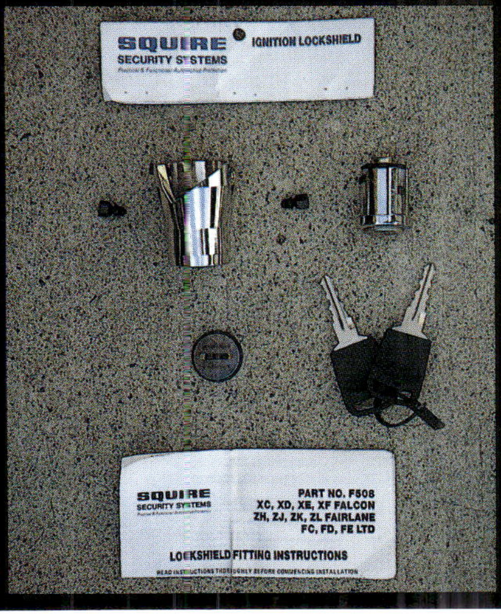

3

4

barrel and cover the entire barrel assembly with a hardened steel cover. Once this unit is fitted it's very hard to remove, even by the owner or a locksmith. We strongly recommend you either take very careful notice of the instructions in the kit or, better still, have it fitted by a locksmith.

Gerard and John of WH Raisbeck and Sons (Locksmiths) kindly agreed to fit the system for us. This company is owned and operated by dedicated street machiners and can supply and fit all manner of car security products.

Give 'em a call on (02) 899 5568 or 899 5569 and ask them how best to protect your pride and joy. They have a variety of different car security products available and can supply and deliver Australiawide as well as advise fitment methods over the phone.

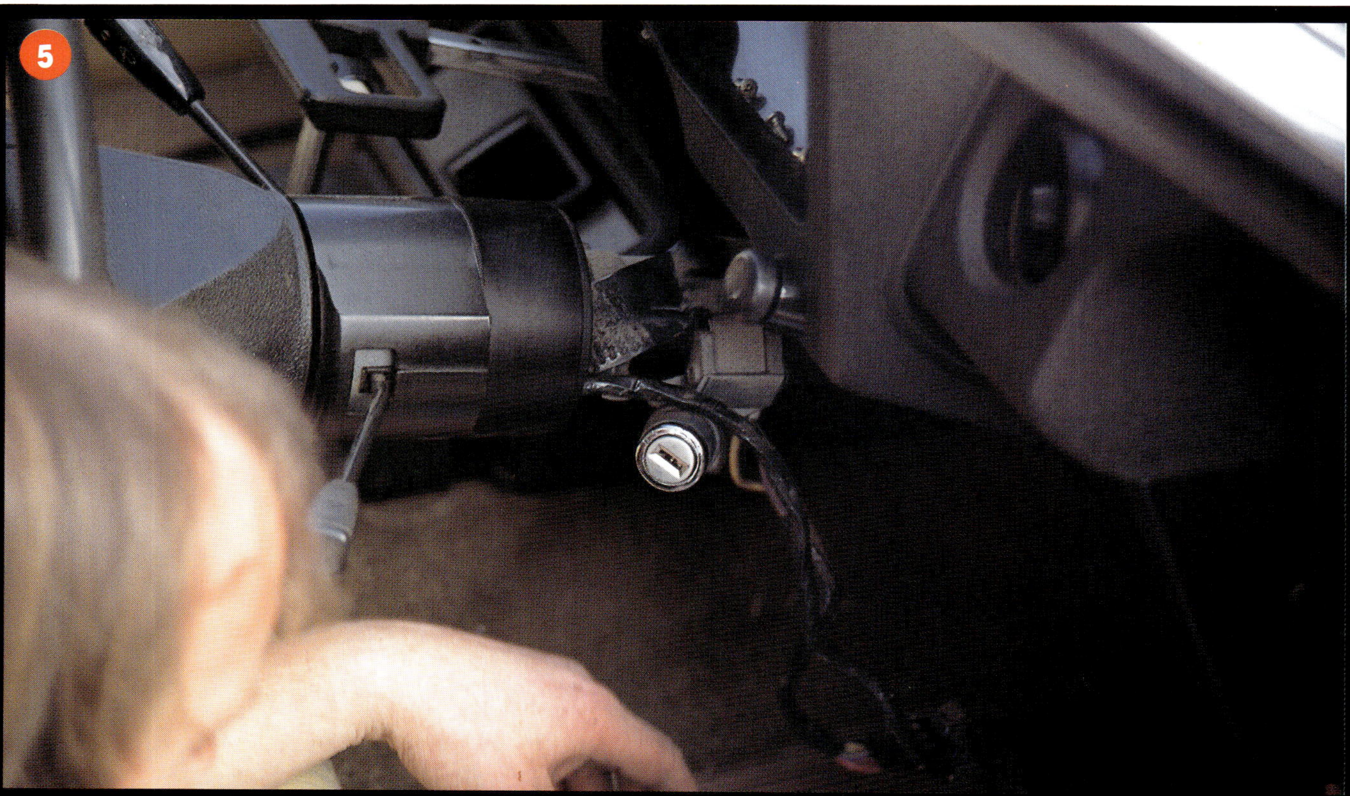

5 The lower column shroud encapsulates the wiring harnesses from the ignition and the indicators. Locate the plugs for these under the dash, unplug them and then unscrew the shroud from the cables.

If your ignition barrel has not been ripped out you need to put your key in, turn it to the on position and insert a small rod or screwdriver into a hole at the base of the lock barrel shroud. The lock barrel can now be pulled out.

6 The replacement barrel is inserted, with the key in place, in the reverse order.

7 The chromed steel shroud and hardened end plate are lubricated with the supplied white grease.

8 With the end plate inserted into the shroud, the entire shroud assembly is then placed over the ignition lock assembly and bolted in place with the break-off bolts. These are designed to lock the shroud in place and then break off at the head, making removal very difficult.

9 The fitted lock and shroud doesn't look much different from the original but protrudes slightly further out of the column shroud.

10 Refitting the column shroud may require, as listed in the instructions, the enlarging of the hole for the lock assembly by a millimetre or two. Mine did not need this but it was a really tight fit to replace.

Be sure to relocate the wiring in the shroud before attempting this to ensure the correct fit. Reconnect the ignition and indicator plugs under the dash at this time as it is easier with the column in the dropped position.

11 Before bolting up the column again, test to see that the car will start and the indicators, horn and high beam switch are working. If all is in order, carefully align the column support bolts, lift the column and bolt back in place.

12 The finished job, as mentioned earlier, does not look that different from the standard setup, but a professional thief will instantly realise that this type of lock is nearly impossible to rip out and he will go hunting elsewhere.

Cruise control was once an item only available on imported or luxury cars. It was often dismissed as 'just another useless addition to justify a higher sticker price', serving no greater purpose than to encourage laziness and pose value. Nothing could be further from the truth.

Wordz by James Garrod **Photos By Mark Hanna**

Cruise Time

If you've ever owned or driven a car with 'cruise', you'll have to admit it's damned hard to live and drive without! A cruise control unit can help you achieve better fuel economy on trips, reduce driver fatigue, make your driving more enjoyable and possibly even save your licence!

I've owned several cars with cruise control fitted as standard or optional equipment, among them Mazdas and Fords, but the best was my VH Brockadore. What if your car doesn't have cruise or you can't find a wreck that has it fitted? Easy. You can buy aftermarket cruise control kits to fit your car. But be warned, you get what you pay for!

As with many aftermarket accessories, there are several levels of quality and features and, the more you pay, the better they get! The Light On brand of cruise control unit is made by an OEM, or original equipment manufacturer. This assures you of quality manufacture and guaranteed parts and service backup. A full range of accessories and fitting kits means the 'Light On' brand can be fitted to any vehicle — auto or manual, diesel, petrol or LPG — and a full parts and service backup is available.

While we're quite capable when it comes to most mechanical tasks, and you've seen our greasy hands in more than one tech feature over the years, it's hard to work and photograph at the same time, so we arranged for Nigel of Cruise Control Specialists in Sydney to fit our unit for us. Cruise Control Specialists is an aftermarket accessory company in NSW that offers a mobile fitment and repair service for this type of equipment.

The Command Auto Group of Botany in NSW was kind enough to provide a unit for fitment and evaluation.

Vacuum hose fitted to actuator unit and mounting bracket is turned around for fitment in the factory (V6) position.

Actuator unit mounts below airbox in V6s and in front of strut tower in V8s.

On VR/VS models there are two rubber hoses located between the airbox and inner guard. One hose carries the washer water and the other is a vacuum hose from the manifold to a vacuum tank in the guard. This tank stores vacuum to operate the heater controls. VN/VP models have only the vacuum hose.

Nigel checks under the guard to find the vacuum hose. "By the way, your washer bottle needs filling," he told us. When you determine the correct hose, trace it back up into the engine bay, cut it and fit the supplied T-piece.

Remove the plenum cover, on VS models, and you will find a mounting point for the actuator cable. For other cars you can use the supplied bracket but you can easily order the factory bracket for a heater installation. Start engine, adjust any slack from the cable and check for smooth operation of your accelerator cable (no sticking or higher idling).

Remove lower dash panel, side vent connector and tray above pedals for under-dash access.

CM2 control unit in place — looks like it was made to suit this hole.

Mount the 'Brain' module in a convenient place. Nigel mounted this unit to the under-dash crossbrace.

CRUISE CONTROLS EXPLAINED

Cruise control is a system that virtually takes control of your accelerator or throttle while you are cruising on open roads. It will maintain the speed you set into the unit to within a few kilometres. The set speed is maintained regardless of whether you are driving up or down hill. The cruise control monitors your car's speed four or five times a second and minutely adjusts the throttle (speed) accordingly. By doing this

Our ute is not fitted with the factory speed alert feature and has a small coin tray in the dash instead. The CM2 controller fits this hole exactly but the On/Off switch can't be accessed. We chose to leave it turned on permanently and fit the controller here. Nigel drilled a quarter-inch hole for the wiring harness, passed the wires through and fitted the controller. Note the lousy fit of the column shroud — it's been like that fom new!

For an auto trans car, like ours, remove the yellow tach sensor wire and the purple clutch sensor wire from the harness.

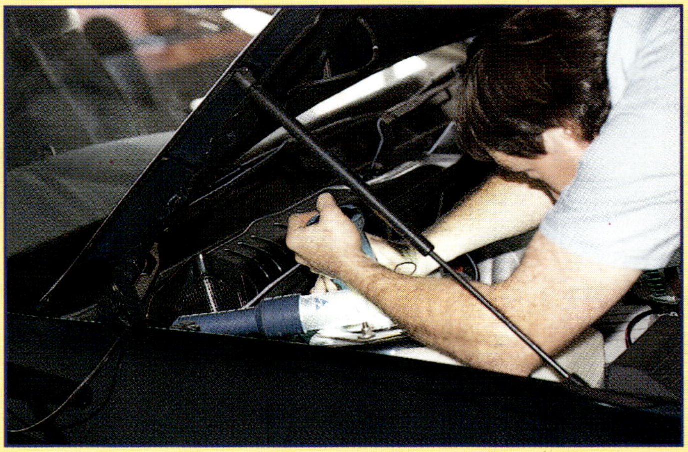

Nigel drilled a hole through the firewall, beside the booster near the accelerator cable for the wiring harness to come through. You can do this yourself but if your Commodore is a manual unit you can pass the wiring through the clutch cable grommet.

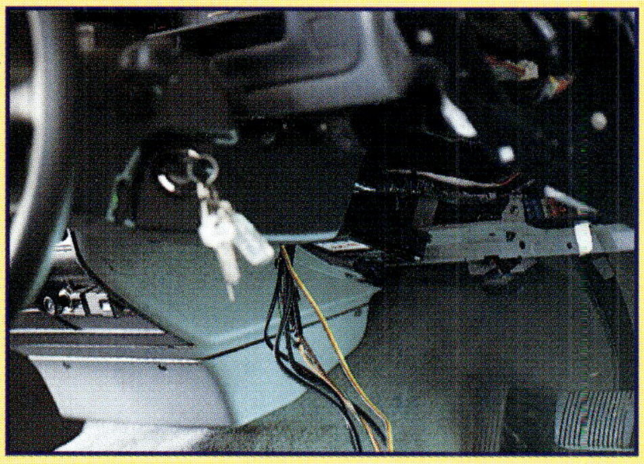

Run the wiring harness under the steering column, not over it, as it can be damaged by the steering uni. Determine which wire goes where, according to the instructions, and trim to length.

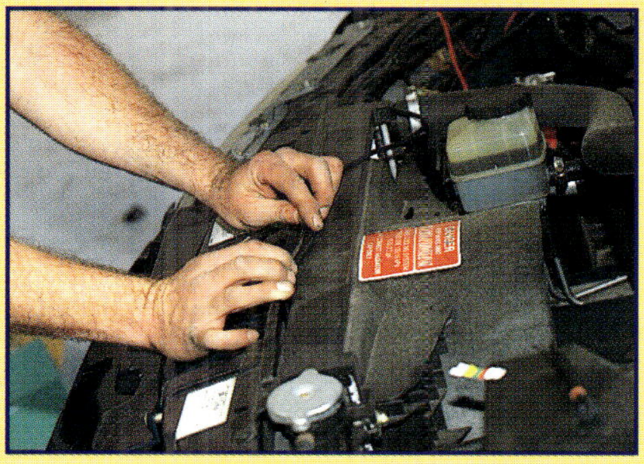

Run the wires for the actuator down the driver's side of the engine bay, next to the inner guard, and across the front of the radiator, down to the actuator unit, following the factory wiring harness. Secure with the supplied wire ties.

Connect the pair of brown wires to the brake-light switch wires and connect the orange wire to accessory power at the fusebox. Wiring connectors are supplied but Nigel solders every connection for a better contact and a quality job.

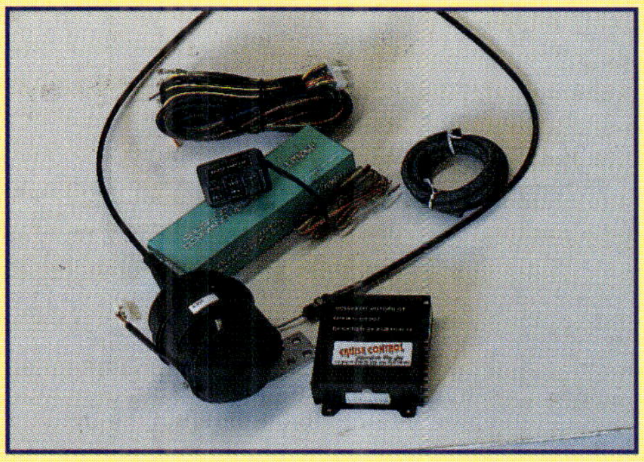

The contents of the kit — everything you could need and then some!

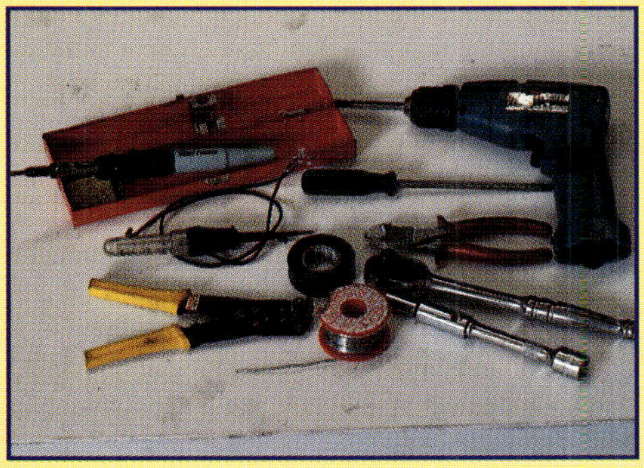

The basic tools needed for the job: soldering iron, cutters, wire strippers, screwdrivers and drill.

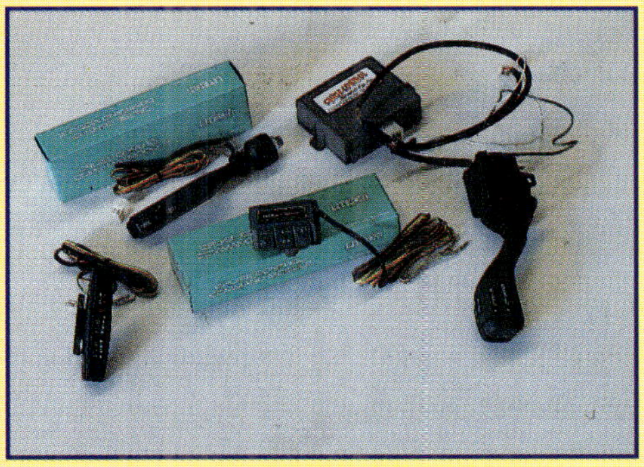

The alternative control switches that are available for different applications.

automatically it improves your fuel economy, because the driver is the greatest fuel-wasting element of any car. Light On cruise control systems have several built-in controls and safety devices that operate, reset or cancel the units.

To operate the units, the main switch is turned on. It will remain on until the ignition is turned off. Once you accelerate to a desired speed, from 40-50km/h up, and press the 'set' button, the car will maintain the set speed until you do one of several things. If you wish to overtake or accelerate, simply press your accelerator pedal as normal. When you remove your foot the car will decelerate to the preset speed.

There are several other safety features built in to prevent overrevving if the car is placed into neutral gear, for example.

If needed, the Lite On units have a Tap Up and Tap Down feature whereby you can increase or decrease your speed setting in 1km/h increments. To increase speed you can hold the set button on and the vehicle will accelerate until the button is released or it will increase by 1km/h each time the set button is pressed. To reduce speed, press the res/dec button and a similar reduction will occur with each press. To restore cruise control operation after braking or gearshifting you simply press the res/dec button.

The system uses electrical impulses from your electronic speedo sender unit but if your car uses a conventional cable speedo drive, you can still fit a cruise control unit using a couple of different methods.

Small magnets can be fitted to your tailshaft, via stainless-steel tie wire (provided) with a pickup unit attached to the car floor, or one of several different sender units can be purchased that attach to your speedo drive cable. A wide variety of control stalks are also available as is the console/dash controller used in this fitment. It's also possible to buy an interface that will allow the Lite On units to operate from a factory cruise control stalk but this can add to the cost.

We fitted a CM300 unit with the CM3 control unit as the Project VS ute is a column-shifted auto and the CM1 control stalk could not be fitted neatly or in a convenient position. Recommended retail price for the CM300 unit, at the time of writing, was $350 and Cruise Control Specialists charged $150 fitting in this case, but this cost can vary with different types of cars.

Follow the step-by-step fitting guide below to learn how to do it yourself, or call Cruise Control Specialists on (0418) 11 6404. For interstate or country enquiries, the same company can freight a unit to you or you can contact the distributors, The Command Auto Group, on (02) 9316 5033 for the name of your nearest stockist/installer.

Smart car builders know how to avoid troubles with the authorities — talk to an engineer before starting your buildup.

REGO

By Paul Tuzson

There's one important piece of paper any smart car builder must have — an automotive engineer's report. They can be costly, but are definitely money well spent, particularly for your peace of mind.

It all comes down to responsibility. If you modify your car and something goes seriously wrong, perhaps causing an accident, even a fatality, someone has to be held responsible. If your modifications are covered by an engineer's report, and a car fault has caused the crash, you can take comfort in the knowledge that the engineer who issued the report will probably bear the responsibility.

When you look at it in those terms, an engineer's report doesn't seem nearly as expensive or as much trouble to obtain.

When choosing an engineer, make sure they're covered by insurance. If you find one who won't discuss the subject of insurance, you should probably look for someone else. In fact, a good engineer will be willing to discuss different ways of doing things, and you may end up with a better buildup as a result.

Getting the right automotive engineer involved in your build before you even start is a very smart move. It's far easier to find out where you're going wrong before you start work than to wait until you've spent your hard-earned on modifications that will only need to be changed later.

It doesn't matter how much experience you have, an engineer will probably find something that's not quite right. This is not nitpicking — remember that once he signs the report he will be held responsible.

Trying to Make Sense

Registration requirements, as with most things in this country, vary from state to state. The Australian Design Rules constitute an attempt to unify the requirements and, while they apply to all cars, state requirements still play a part. One engineer I spoke to said that he doesn't expect to see uniform state regulations for at least another four to five years.

Manufacturers must guarantee that each vehicle produced complies with the Australian Design Rules when it leaves the plant. Once you modify a car the onus is on you to ensure continuing compliance with whatever rules were in effect at the time the car was manufactured.

No matter how much experience you have, you are not recognised as having the skills to certify compliance, so you need an engineer who is acceptable to the authorities in your state. All of the state authorities

It's far easier to find out where you're going wrong before you start any work than to wait until you've spent your hard-earned on modifications that will only need to be changed later.

publish lists of people whose assessments are acceptable.

Although there are specific automotive engineering courses available, most recognised engineers are mechanical engineers who specialise in automotive work.

Automotive engineers oversee the mechanical aspects of changing a car structurally as well as from a performance point of view. Licensed roadworthiness testers fit into the picture by making sure that the vehicle continues to comply with either the manufacturer's specifications or the modified specifications as set out by the engineer.

Jeff Watters, a Melbourne engineer we spoke to, pointed out that roadworthiness inspectors usually have a very good understanding of the day-to-day practicalities involved.

– THE RIGHT WAY

Queensland has introduced a system which greatly reduces the need to consult an engineer. There's a state-wide list of specialists who are authorised to conduct inspections and certify compliance within their area of specialisation. I asked Queensland Transport for a copy of those on the list and was sent a package so large I had to go down to the post office to collect it.

Authorised officers can certify modifications such as those to the engine, rear axle, brake system and seat substitutions as well as body modifications and conversions and a whole host of other modifications.

Often the seller of a particular modification will also be authorised to inspect it, saving enormous trouble all round. It would be

MODIFICATION LEVELS

Not all modifications need to be covered by an engineer's report. Basically there are three levels of modification:

1. Changing production level. This is when you upgrade a vehicle to include manufacturer's options available as factory fitments for the same model. This generally won't affect registration or street legality, but your insurance company will almost certainly want to know about it.
2. Modifications which are well within the guidelines published by the registration authorities in each state. In Victoria, for instance, you can fit wheels one inch wider than the widest offered by the original manufacturer.

On a Falcon hardtop, for example, the widest wheel size Ford offered as a factory fitting was eight inches. If you bought an old Falcon GS which had six-inch wheels all round and for some strange reason you wanted to replace them with seven-inch wheels, you would be well within the guidelines, and the services of an engineer would be unnecessary.

3. Modifications which are close to or exceed the published guidelines for your state are more usually the case. This involves fitting equipment which was never offered as a factory option for your model car. Most street modifications fall into this category — fitting wide wheels, increasing engine power beyond 10 percent (in Victoria), changing engines, transmissions, brakes, steering etc. In short, almost anything worth doing to a serious car needs to be covered by an engineer's certificate.

Going one step further and pushing your car into the highly modified bracket leaves no doubt at all — you will definitely need an engineer. Highly modified doesn't really need a definition; it means just what it says.

sensible if all states adopted such a system. Doing things this way wouldn't put engineers out of business as there are still many modifications which are outside the system.

If a modification does fall outside the Authorised Officer scheme, you will need to submit a detailed plan to Queensland Transport before commencing work. It's up to you whether or not you supply an engineer's assessment of the project at the preliminary stage, but doing so will ensure that your plans don't contain ideas which will be instantly rejected. When the project is finished it is presented to Queensland Transport for inspection. An engineer's report is not necessary at this stage.

In Western Australia, the Police Licensing Branch has the final say. If the modification is from the original vehicle manufacturer or is a fairly minor one, the issue of a permit is straightforward. If the modification is more substantial, you must submit a detailed proposal which will be accepted, rejected in part or completely rejected.

Assuming your proposal is accepted, it's up to you to get the job done properly and then present the car for inspection. Engineer's certificates play a role but they are not the last word — the Police Licensing Branch officers have the ultimate power.

If the police pull you over, you need only produce your copy of the report to have the matter resolved on the spot.

In New South Wales, advance permits for heavily modified vehicles are not needed and such vehicles are no longer inspected by the RTA. Instead, you must take it along to an Authorised Unregistered Vehicle Inspection Station (AUVIS). They will decide whether it is acceptable and issue a change of records notification. An engineer's certificate is needed.

For modified vehicles in Victoria the RTA says it needs engineering proof from somewhere. If you provide a report from one of their recognised engineering signatories

The red two-door XP shown was engineered by Jeff Watters. It's fitted with a 351 V8 which has had all sorts go-fast work done to it. As the photographs show, the pillars had to be cut back in order to fit the motor. Any body cutting like this should be covered by a report, no matter how much you know about cars.

they say that the vehicle will be automatically accepted. If you don't provide a report from a recognised source, they will check the vehicle out themselves.

Because they are geared to rely on recognised signatories, organising for the RTA to perform the check could take quite a bit of time. If they decide they want to look at something which should have been inspected before the car was finished, you could be asked to do all sorts of inconvenient things like removing carpets etc to let them get at the modified area.

South Australia says that very minor modifications don't need an engineer's report. But once again, anything really worth doing does require the involvement of a certified engineer. I spoke to the authorities and they said that they won't look at a heavily modified car unless it comes to them with a report.

However, as in Western Australia, an engineer's report is still no guarantee that the car will pass. The Vehicle Engineers Section checks the vehicle and has the final word. Obviously local engineers know what will and won't be accepted, so in theory you should be able to get your car registered if you have a certificate.

Costing the Certificates

The cost of an automotive engineer's report varies according to the amount of time involved in assessing your project. Naturally, the more complicated your build, the more things will have to be checked and the more it will cost.

Yet even the comprehensive projects shown in this article didn't cost more than about $400 to certify. Engineers set their own prices and there can be large variations from one to another for the very same project. I heard of one engineer who quoted over $900, but another charged only $250 to check the same mods on the same car.

For your money, you get three copies of the final report and the peace of mind that your car really is safe. One copy goes to the authorities, one stays with the engineer and you get the third. You should keep it in the car.

In addition to ensuring that you will get your car registered and insured, this also buys you hassle-free driving. If the police pull you over, you need only produce your copy of the report to have the matter resolved on the spot — but remember, the police still have tremendous power and can put you off the road if they don't like your attitude.

If things did go that far (and they never should), the report will almost definitely clear the matter up when the car is presented for inspection.

There is, though, one situation that really annoys engineers and puts owners in breach of the law — cars which are presented for final inspection in one condition then modified again later without additional consultation. Doing this will probably invalidate any insurance you've been able to get. Plan ahead and build it right the first time.

BEGINNER'S GUIDE TO mig welding

Learn how to MIG weld and you will get a great deal of pride from your work as well as saving plenty of cash, and lots of time.

BY PETE 'SPANNERMAN' DUNCAN

WELDING appears easy at first glance but when it comes to the crunch it ain't. Just ask any first timer who rather than welding two pieces together ends up with a huge hole where some important metal used to be!

Then, of course, there is the jargon of welding, with names such as oxy, MIG and TIG tripping off the tongues of those in the know (and often from the tongues of those who haven't got a clue!).

Let's face it, most people don't know how to weld and are often too scared to give it a go. As a consequence, small tasks are sent out for professional attention and are returned with a huge bill attached.

If you intend building a car you really should know how to weld not only as a means of cutting costs but for the convenience factor. And your mates will love you all the more once they discover you can make a welder work.

Of the various types of welding styles, MIG is the simplest and the basics can be learnt in about 90 minutes by a total novice. It may well be the best tool you'll ever buy. Once you buy a MIG you'll wonder how you ever managed without it.

How It Works

MIG stands for Metal Inert Gas and is a semi-automatic welding process which combines electric arc welding with gas shielding. Its popularity has gained a lot of momentum over the years, particularly in the panelbeating trade, because the fast welding speed helps keep panel distortion down.

A MIG welder feeds a spool of wire through a watercooled gun with an inert gas shield (usually argon) which provides the same protection as a flux, i.e, it reduces the tendency of the metal to oxidise when heated.

MIG welding is two to three times as fast as normal gas welding and the operator doesn't have to clean flux or slag off the welded surfaces afterwards.

MIG welding is claimed to be as strong as welding by conventional systems. It is done at temperatures of about 6000 degrees C, the same temperature as electric welding (oxy-acetylene is done at about 3000 degrees Celsius), however the equipment is usually dearer than that for most other welding processes.

Two pieces of metal can be joined if they are heated to the point where they locally melt, flow together and then cool off to form as one. MIG works by supplying heat, in the form of electricity, to the metal that is to be welded, through a wire fed through the tip of the torch.

A circuit is made with a large earthing crocodile clip fixed to the work so that when the welding wire touches the work electricity flows and very quickly heats up the area the torch has touched.

Carbon dioxide (CO_2) gas is introduced at the same time as the torch's trigger is pulled, and the wire fed. That's to stop the weld oxidising, which would otherwise make it porous and weak. The bonus of this type of welding over gas or arc welding is the localised heat. That means reduced distortion, plus the fact that there is no naked flame involved, so it's also safer than gas welding.

Aussie Made

We called on the services of the Croydon, Melbourne-based Ausmig company for help with this feature. From what we could decipher, Ausmig is the only major Australian portable welding equipment manufacturer. Ausmig says that, while others exist, none has Advance Australia Foundation Certification, meaning that a minimum 75 per cent content must of Australian origin.

We found the product to be of the highest standard, incorporating features such as an automatic wire speed and a voltage synchronisation system. Ausmig produces its own copper transformers, it doesn't use aluminium as is found on some imports and which Ausmig says is inferior. The company offers a four-year warranty on this component.

Features abound, and the extensive range includes 240-,415- and 480-volt models and a choice of 35 per cent, 60 per cent and 100 per cent duty cycles. Pricing kicks off at $1890 with the monster Terminator A500T topping out at $5550.

Ausmig also offers an extensive range of accessories and has recently introduced its own range of plasma cutters. All are available through distributors in every Australian state.

Make sure you check out the Australian-made Ausmig you won't be disappointed. If this sounds like we're pissing in their pocket it's because we took the time to check out some of the cheap imports and quickly came to the conclusion that you only get what you pay for.

4. The wire feed assembly consists of the wire roll (sizing depends on model) and automatically controlled wire-drive tensioner which works in conjunction with voltage synchronisation system.

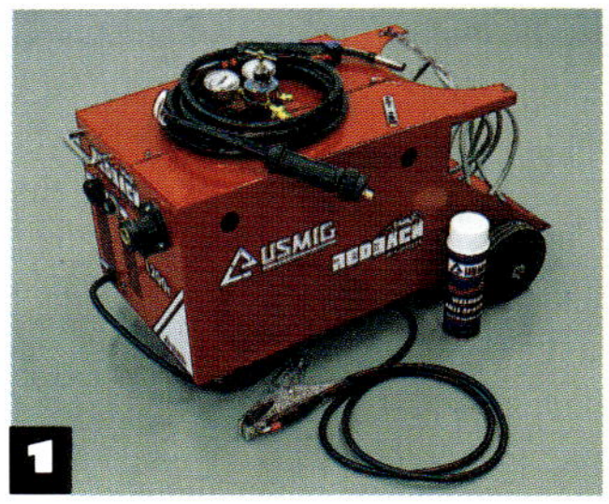

1. Ausmig units come complete with everything you need except gas bottle.

2. Once the wire spool is placed in position it is run through the opened wire-drive tensioner, ensuring the appropriate roller is used with the selected wire. Once approximately 30mm of wire is fed through the guide tube the tensioner can be closed, the unit switched on and the wire speed dialed up to start the wire drive rollers. Trigger remained depressed until the wire protrudes from the gun head.

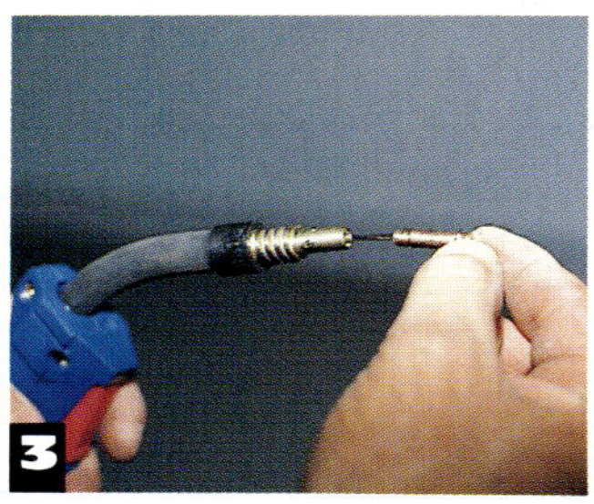

3. With the wire protruding, the appropriate tip can be fitted. The gun assembly differs depending on what is to be welded, be it steel, stainless steel, aluminium or non-ferrous metal.

5. A twin-gauge regulator is supplied for the gas bottle. It consists of one gauge for bottle pressure (remaining gas) and one advising the flow rate. The flow rate is adjusted as desired via a control knob.

6. The power of the welder is adjustable so you can weld different thicknesses of steel. All Ausmig units have two knobs which, on the smaller commercial models, offer manual and synchronised adjustment. The thicker the metal the more power and the quicker the wire feed needs to be. Every MIG comes with an easy guide that lists metal thicknesses against the required MIG settings.

7. When you weld the torch should be held approximately at this angle and distance (around 10-15mm) away from the material. For a good weld it is important that you maintain the same distance away from the steel otherwise it will splatter.

8. You must wear protective clothing including mask, gloves and overalls.

Before trying to weld you need to fine tune the wire feed. Take a piece of scrap metal, turn the gas on, and refer to the MIG's manual for power settings. Pull the trigger and start to weld. Listen to what's happening, noting that a sound like frying bacon means that wire feed is either too slow

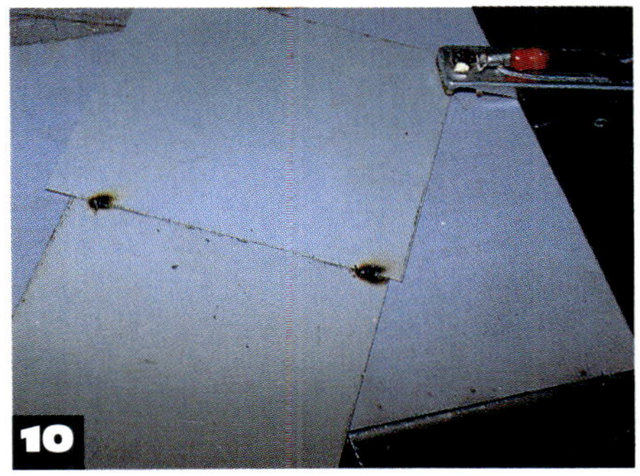

10. The first technique is butt welding where two pieces of metal butt against each other. Tack a blob of weld at each end, or at regular intervals, bearing in mind it should only take a second or two for each tack. You can then either weld up the gap in one long weld, slowly moving the gun along without stopping, the other option is to...

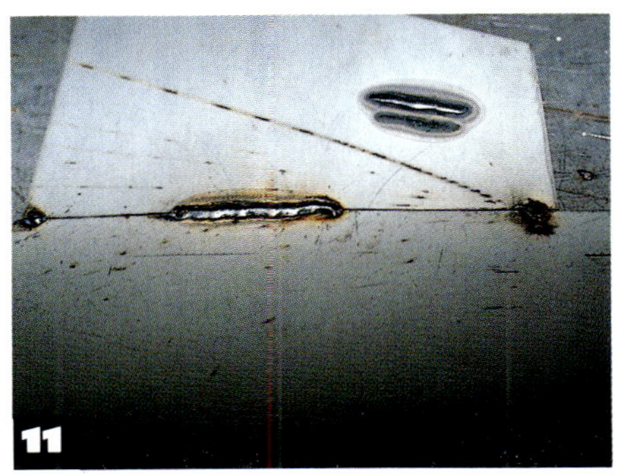

11. ...when welding tin panels, press trigger and weld for a couple of seconds until the weld is quite red, stop for half a second while not moving the gun before

In other words, short bursts are followed by very short stops. When welding thin material, if it's allowed to get too hot there is a chance it will warp or burn holes.

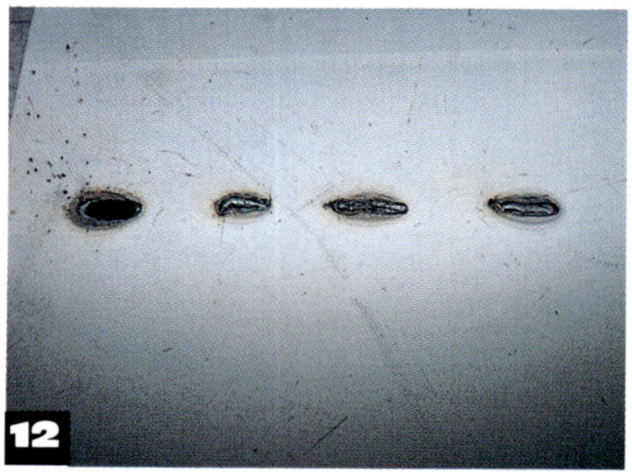

12. If you set the power too high on the welder, or the metal was rusty, the results will look like the example on the left. If the power was too low a joint may have been formed but the weld will sit on top of the metal, (second from the left), rather than sink into it as it has on the far right example. Second from right shows a hollow weld which is probably caused by a low gas supply.

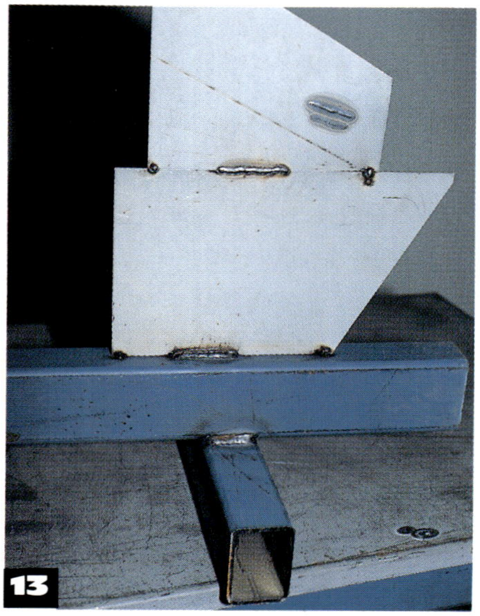

13. Once skilled you will be able to easily adjust machine to suit varying metal thickness — as depicted

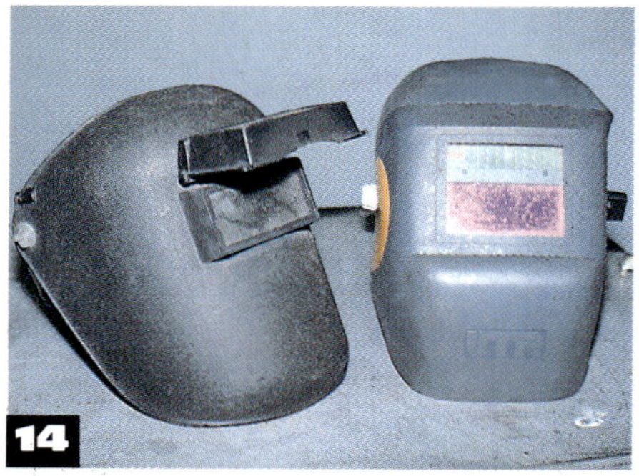

14. Two types of masks are available, the traditional mask which requires a visor be lifted to allow placement of torch. The second, is a UTP version that automatically adjusts (darkens) once the torch starts welding. By the way, the extremely useful UTP mask costs somewhere around $2500!

MADE IN JAPAN

CHEAP THRILLS

Pssst. Want to buy a fuel-injected, quad-cam, 204kW V8 for a couple of grand? Not too choosey where it comes from? Nippon down to your local wrecker and witness the future

ONCE upon a time, Japanese factories churned out four-cylinder buzz boxes and little else, but in the past decade they've learned how to build a host of strong, smooth turbo-charged sixes and free-breathing, high-tech eights. A good example is the all-alloy Lexus V8, which produces some 273hp (204kW) straight out of the factory door!

But a Lexus LS400 costs something like $150,000, so who cares? Well, what if we told you that you could pick up the whole front half of a Lexus — motor, four-speed auto trans, wiring, computers, brakes, dash — for about five grand? Or the engine on its own for only a couple of thou? We're talking six-bolt mains, ARP-style rod bolts, double-overhead cams per bank, the works. Change the restrictive exhaust manifolds alone and add another 30hp. And wouldn't a quad-cam V8 look good under the bonnet of your Commodore? Would it what!

But while the Lexus V8 is the best-known of the big-bore, hi-po Japanese engines available secondhand at your local Japanese engine importer, it's definitely not alone. How's this lot for starters?

Story & photos by Gowrey

Nissan

VG30DETT twin-turbo three-litre V6

TRANSLATED, the Nissan VG30DETT donk is a three-litre V6 with twin turbos. Don't worry about the fact that it's only three litres — it still pumps out 280hp at 6400rpm! You'll see that 280hp figure fairly often in this feature, because it's the Japanese legal limit. But in real life, lots of the big Jap engines produce more power than this figure — especially the turbos, and especially if you tweak 'em.

The Nissan VG30DETT is a twin-turbo, three-litre V6 powerhouse. On this crate is the full engine wiring loom, all the sensors and the computer. Try driving one without it!

This motor is fitted in Japan to the Nissan 300ZX Turbo (same shape as the last Aussie-delivered 300ZXs). It's a pretty big car, and needs to be because the engine is 700mm wide, 550mm long and 620mm high!

The sticking point in fitting this engine into a new body is finding the room for the exhaust pipes off the turbos. These pipes can get real close to fouling the firewall in a heap of cars, so it will depend to some extent on how far forward you mount the engine. The other thing to keep in mind is that accessories like the air-con compressor are mounted relatively low, making the whole assembly wider at the bottom than you'd otherwise expect.

When doing your measuring, keep in mind that the factory engine runs two intercoolers at the front of the car, one on each side. You could replace that arrangement with a single large 'cooler or keep the twin system — but either way, you'll have to find room for at least something.

In the States, there's a whole heap of go-fast goodies for this engine; getting 450hp out of them is pretty easy, in fact. Parts are available from Nissan for the naturally-aspirated VG30DE, but not for the twin turbo unit which was never sold new here.

Because the naturally aspirated VG30DE was fitted to Australian-delivered 300ZXs, some parts are available that also fit the turbo engine. Note how the engine really fills the big bay, even in what is a pretty big car.

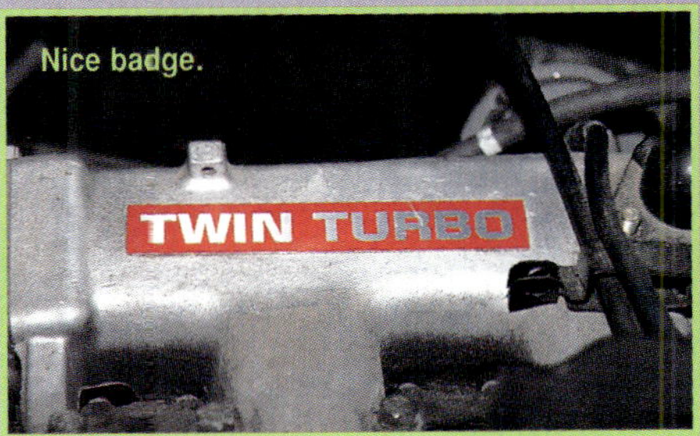

Nice badge.

RB25DET & RB26DETT straight-six turbos

OK, now listen up real close. These turbocharged DOHC Japanese-delivered straight sixes, with either 250hp or 280hp on tap, will bolt straight into your VL Commodore!

Why? Here's a quick history lesson. When Holden had to switch their engines to unleaded in 1986, they searched the world for a new engine. They didn't have the bucks to immediately develop their own, and they ended up picking Nissan as the VL six engine supplier. Nissan made available two sixes — the RB30E three-litre atmo motor, and the RB30ET turbo version of the same engine. Both were single-cam 12-valve engines. The turbo version was developed just for Holden and wasn't used in Japan by Nissan. Huh? The answer is Nissan had plenty of better engines to play with!

Engines like the 250hp DOHC 2.5-litre turbo (the RB25DET) and then, later, the stomping twin-turbo 2.6-litre 280hp RB26DETT, the same motor that powered Godzilla, the Bathurst-winning Nissan Skyline GT-R! Yep, the Australian VL six is a very close relative of the GT-R Bathurst winner's engine.

The fact that all these engines have code numbers starting with 'RB' indicates that they're from the same family, sharing the same blocks. Too easy!

There are two approaches you can take in getting one of these superb, smooth and grunty turbo sixes into your Commodore. First up, a straight engine swap. If you've got a VL six you can pull out the old engine and plonk onto the same mounts the 250hp RB25DET. You will need the new engine management computer and loom, and investing in the complementary gearbox will make things a bit more reliable.

If you have much deeper pockets, you could do the same thing with the twin-turbo RB26DETT GT-R engine, although you'd have to buy a gearbox and diff as well to take it: even though it's rated at 'only' 280hp, you can expect it to have up to 320hp once it's sorted.

Or, if you want to go the full Monty, you can use the twin-cam head from either of these engines on the Australian-delivered three-litre block! If you want to go absolutely apeshit, you can even bore the block out to 3.1 litres, bolt on the RB26 head (complete with solid tappets and 8000rpm redline), and then fit the twin turbos! It's been done, and the result is about 650hp of streetable muscle! And that ain't no misprint, either.

Of course, for this sort of development, you need fully programmable management (say, Autronic or MoTeC) and a very experienced engine builder who knows how to read a dynamometer. Both AVO and Nizpro in Melbourne can advise you on this swap.

Parts are available from Nissan for the RB26DETT engine because the Skyline GT-R was sold here in small numbers. The RB25DET was not available here new and so Nissan don't stock parts, but we hear rumours that bits like head gaskets can be ordered through Nissan without much problem.

The Godzilla Skyline twin-turbo six will bolt straight into your VL Commodore, giving you at least 280hp, and up to 350hp if you try.

VH45DE & VH41DE quad-cam V8s

WHEN Lexus produced the LS400, Nissan produced the Infiniti Q45. Never heard of it? You're not alone. The Infiniti (a large luxury sedan) was sold in very small numbers in Australia, but the good news is Nissan stocks parts for its engine, an all-alloy DOHC V8, the VH45DE. Punching out 270hp at 5600rpm, the 4.5-litre engine is the largest of the new-generation Japanese V8s.

What is even less well known is that the VH45DE has a smaller brother, the 4.1-litre VH41DE. Showing how the Japanese power figures are a bit rigged on the low side to cheat the local authorities, this engine also claims 270hp, although in this case at 6000rpm.

Both engines are extremely rare in Australian Jap-importing wreckers, but we have started to see a trickle of VH45s. If you find a VH41, ask for a discount after you point out that it's just the little engine. (Don't tell them we sent you.)

Toyota

1UZ-FE four-litre V8

THIS is the Lexus engine that grunts out between 260 and 273hp, depending on how recent it is. Becoming a favourite in the States where some astronomical power outputs are being got out of blown versions, the engine is extremely strong and beautifully made.

How beautifully made? Well, what if we told you the pistons, blocks and even the head gaskets are all matched to each other on the production line? That one LS400 has done more than 330,000km (so far) with nothing going wrong at all? That the biggest job my local Lexus dealer has EVER done is to replace a rear main seal? One long-time Australian drag racer described its bottom end as the best that he has ever seen in a production engine.

We don't want to offend anyone, but this engine (and the others discussed here) was designed something like 30 years after most of the V8s seen in Australia. And it shows. Take the oil pick-up, for example. Compared with the

"We don't want to offend anyone, but these engines were designed 30 years after most of the V8s you see in Australia...and it shows"

Fitting twin-turbo exhausts is often a problem when installing the VG30DETT into a body not originally designed for it. They can hit the firewall if you're not careful.

Empty...

Holden V8, the pick-up filter is huge, as is the feed line to the pump. The cast-alloy sump uses special guides within it to channel oil back to the pick-up, and to prevent oil surge. That sort of thoroughness of design can be seen right through the engine.

Across the cam covers, the engine is 680mm wide and 670mm long — just about square when you look at it. From the very bottom of the sump to the top of the intake plenum it stands 700mm high — so let's call it a cube!

As with most of these Jap engines, you'll find a four-speed auto trans backing the engine. The Lexus 'box is fully electronic, which makes it easy to dial in if you have the factory computer. Even better, early investigations show the electrical resistance of the solenoids and their functions are the same as the Commodore four-speed auto, so it may be possible to run it using the Powertrain Control Module from late Commodores. Cool! Some recalibration of the Holden Mem-Cal would be needed because the gear ratios in the Lexus box are different from the Holden auto trans, but you get my drift.

Incidentally, the Lexus gearbox is smaller than the Commodore unit and uses six bolts to hold the torque converter to the flywheel, rather than the three or four bolts used in many other makes.

While you can get DOHC conversions for some pushrod V8s, why would you bother when engines like this can be picked up for a few grand?

...and full of Lexus V8!

Because this engine is fitted to the Australian-delivered Lexus LS400, parts are available locally for the engine and gearbox.

7M-GTE turbo six

THE 7M-GTE (and 7M-GTEU) is a DOHC single-turbo, three-litre straight six, the engine you'll find in the Australian-delivered Supra Turbo. Grunting out 240hp at 5600rpm, this mill has the advantage that full parts support is available through your local Toyota dealer.

The inter-cooled six is another that responds well to turbo boost, exhaust and intercooler mods — the quite heavy Supra runs mid 13s with this sort of inexpensive work.

However, when producing absolute max power, these engines have been known to spit head gaskets and not be as reliable as some other Jap imports. But if you have your heart set on this engine, you can sure upgrade it — there's whole heap of good stuff available in the US.

At the wreckers, you'll find this engine is not as cheap as others because all the Supra owners who have blown their engines have cornered the market! Bastards!

2JZ-GTE & 1JZ-GTE twin-turbo straight sixes

THESE are pretty recent hi-po Jap engines. The 2JZ-GTE replaced another twin-turbo Toyota six, the popular two-litre 1G-GTEU. That older engine only had 200hp, which is OK but nothing flash for twin overhead cams, 24 valves and two turbos! After all, the single-cam 12-valve VL Turbo engine has the same power...and more torque from its bigger capacity.

Anyway, the latest Toyota twin-turbo sixes are very different engines from the earlier model. The 2JZ-GTE boasts 280hp at 5600rpm from a capacity of three litres. It's fitted to the current shape

The Supra Turbo weighs in at nearly 1600kg but can still turn mid-13s without a lot of work.

CHEAP THRILLS

This is the Lexus V8 engine. The queue starts here.

Six-bolt mains and serious heavy-duty construction...locally delivered cars are racking up 330,000-plus kilometres without problems. And it's no wonder — just look at the size of that oil pick-up!

The 2.5-litre 1JZ-GTE twin-turbo six is externally identical to its bigger three-litre 2JZ-GTE brother. Both engines quote 280hp as standard, but only because it's against the law to admit to any more in Japan.

"One privately imported Soarer, with 15psi boost, a new intercooler and a K&N cone filter, recorded 240hp at the rear wheels"

"I'll have a half." This is what you call a 'half-cut' — you get the complete front half of the car. Buying this 2.5-litre twin-turbo half-cut means you've got a better than even chance the engine will go without spending any more money.

Toyota 7M-GTE turbo sixes were fitted to the Australian-delivered Supra Turbo. That makes the 240hp six a good proposition with spare parts available from local Toyota dealers.

Supra (the one with the bulging headlights) which we haven't seen in Australia, except in small numbers of private imports. But like the other turbo engines discussed here, it responds really well to simple mods that don't require you to get into the engine.

Here's an example. As well as the three-litre version, the engine (complete with its twin turbos) is also available in 2.5-litre capacity (the 1JZ-GTE), and claims the same factory figure of 280hp. (Funny that.) This smaller engine came out in Japan in a luxury coupe called the Soarer. One privately imported Soarer measured 150hp in standard form at the rear wheels. But with 15psi boost, a new intercooler and a K&N cone filter, the power went up to 240hp at the wheels!

See what we mean about responding to minor mods? In the 1800kg Soarer (a bloody heavy car), the quarter-mile blasted past in 13.7 seconds at 110mph. Hey, and it idles! ■

CONTACTS

THE engines that you see here were all photographed at the following businesses:

■ **Awesome Automotive**	**08 8277 3927**
■ **Adelaide Jap Dismantlers**	**08 8369 1156**
■ **Japco Auto Wholesalers**	**08 8347 3599**
■ **Jap-Parts Rosewater**	**08 8341 1600***
■ **Jap-Parts St Marys**	**08 8276 1033**

To buy or not to buy

IF buying from a Jap-importing wrecker, there are three ways you can go about it. First up, buy just a bare engine, maybe with trans attached. Or you can buy the engine and trans with the wiring loom and computer. Finally, you can buy the complete front half of the car — called a 'half-cut.'

All these engines are complex bits of gear, which makes getting one up and running without the loom or computer a big ask. Like, you're going to need to use programmable management (MoTeC, Autronic), which will set the budget back at least another $3000. Not only do you have to buy the fuel and ignition computer, it then needs to be programmed!

If you get the bare engine with loom and computer, make absolutely certain you have every EFI sensor (for example, airflow meters and ignition amplifiers) and that the loom is uncut. Finding an engine with an uncut loom and every sensor intact is pretty bloody rare, to be honest, but worth the effort.

Finally, there's the half-cut approach. This is by far the best way of doing it, because you know that you have everything needed to make the engine go. You also get some bonus bits that you can either sell or use in your project — bits like front brakes (as trick as the motor), air-con, dash gauges, radiator and so on. On the turbo engines, you'll also get the intercooler(s).

So don't just rush into buying a bare engine and trans without carefully considering the flow-on costs to make it all work, which can double the cost of that initial buy. Good luck!